A Guide to the

Wireless Engineering

Body of Knowledge

(WEBOK)

2009 Edition

A Guide to the

Wireless Engineering
Body of Knowledge
(WEBOK)

2009 Edition

G. Giannattasio, J. Erfanian, K. D. Wong, P. Wills, H. Nguyen, T. Croda,
K. Rauscher, X. Fernando, N. Pavlidou

IEEE COMMUNICATIONS SOCIETY

IEEE Press

A JOHN WILEY & SONS, INC., PUBLICATION

Published by John Wiley & Sons, Inc., Hoboken, New Jersey. All rights reserved.
Published simultaneously in Canada.

For general information on our other products and services please contact our Customer Care Department within the United States at (800) 762-2974, outside the United States at (317) 572-3993 or fax (317) 572-4002.

Wiley also publishes its books in a variety of electronic formats. Some content that appears in print, however, may not be available in electronic formats. For more information about Wiley products, visit our web site at www.wiley.com.

Library of Congress Cataloging-in-Publication Data is available.

ISBN 978-0470-43366-9

Printed in the United States of America.

10 9 8 7 6 5 4 3

Contributing Authors

The following is a list of volunteers who have contributed to the writing, editing, and reviewing of the 2009 Edition of the Guide to the Wireless Engineering Body of Knowledge. The WEBOK would never have been created without the support and participation of these volunteers. IEEE and IEEE Communications Society would like to thank and acknowledge them for their selfless contributions.

WEBOK

Editor in Chief	**Gustavo Giannattasio**

Chapter 1 Wireless Access Technologies

Editor	**Javan Erfanian**
Contributing Author	Rémi Thomas – France Telecom Orange
Contributing Author	Anne Daviaud – France Telecom R&D
Contributing Author	Jin Yang – Verizon Wireless
Contributing Author	Javan Erfanian – Bell Canada
Contributing Author	Paul Eichorn – Bell Canada
Contributing Author	Haseeb Akhtar – Nortel Networks
Contributing Author	Angeliki Alexiou – Alcatel Lucent Labs UK

Chapter 2 Network and Services Architecture

Editor	**K. Daniel Wong**
Contributing Author	Dharma Agrawal – Cincinnati University
Contributing Author	Javan Erfanian – Bell Canada
Contributing Author	Vijay Varma – Telcordia
Contributing Author	Hung-Yu Wei – National Taiwan University
Contributing Author	K. Daniel Wong – Malaysia University of Science and Technology
Contributing Author	Qinqing Zhang – Milton S. Eisenhower Research Center, Johns Hopkins University, Applied Physics Laboratory

Chapter 3 Network Management and Wireless Security

Editor	**Peter Wills**
Contributing Author	Bernard Colbert – Deakin University
Contributing Author	Paul Kubik – Telstra
Contributing Author	Santiago Paz – Ort University
Contributing Author	Peter Wills – Telstra

Chapter 4 Propagation and Antennas

Editor	**Hung Nguyen**
Contributing Author	John Beggs – The Aerospace Corporation
Contributing Author	Asha Mehrotra – The Aerospace Corporation
Contributing Author	Hung Nguyen – The Aerospace Corporation
Contributing Author	Dennis Sweeney – The Aerospace Corporation

Chapter 5   Facilities and Wireless Infrastructure

Editor	**Thomas Croda**
Contributing Author	Filomena Citarella
Contributing Author	Richard Chadwick
Contributing Author	Thomas Croda
Contributing Author	Rolf Frantz
Contributing Author	K. Raghunandan
Contributing Author	K. Daniel Wong

Chapter 6   Agreements, Standards, Policies and Regulations

Editor & Author	**Karl F. Rauscher**

Chapter 7   Wireless Engineering Fundamentals

Editor	**Xavier Fernando**
Editor	**Niovi Pavlidou**
Contributing Author	Anurag Bhargava – Ericsson
Contributing Author	Joseph Bocuzzi – Broadcom Corporation
Contributing Author	Naveen Chilamkurti – La Trobe University
Contributing Author	Xavier Fernando – Ryerson University
Contributing Author	Ali Grami – University of Ontario Institute of Technology
Contributing Author	Stylianos Karapantazis – Department of Electrical and Computer Engineering, Aristotle University of Thessaloniki
Contributing Author	Wookwon Lee – Dept. of Electrical and Computer Engineering, Gannon University
Contributing Author	Evangelos Papapetrou – Dept. of Computer Science, University of Ioannina
Contributing Author	Xianbin Wang – University of Western Ontario
Contributing Author	Traianos V. Yioultsis – Department of Electrical and Computer Engineering, Aristotle University of Thessalonik

Introduction

Wireless technology has provided connectivity and communications for well over a century, providing consumers with previously unknown flexibility and mobility. Wireless coexists with, extends, and even competes with wired communication links. In recent years, the role of wireless technology has broadened significantly, and to serve an increasingly mobile society wireless will need to grow many times over in the years ahead.

The total knowledge dealing with the many aspects of wireless technology will grow accordingly. This Guide to the Wireless Engineering Body of Knowledge (WEBOK) outlines the technical areas with which practitioners should be familiar, and offers suggestions for further information and study

Fundamentally, wireless communication technology depends upon generic communication system principles, and yet, it has its own unique attributes. These include:

- radio engineering
- wireless link design
- the wireless infrastructure
- spectrum and frequency allocations
- networking and mobility management
- services
- user devices and interfaces
- regulatory and compatibility requirements

The goal of any communication system is to connect and transmit between two or more points, be they persons, premises, or machines. A layered architecture stitches together the applications and user interactions, which are being met by increasingly uniform services and service delivery architectures.

A broad range of services exists and continues to grow, enabled by wireless networks, be they fixed or mobile, satellite or terrestrial, conversational or interactive.

The primary mobile communication service has been the voice call, enabled by cellular systems that have traditionally been circuit-switched and optimized for voice. Mobile data services have, however, grown significantly so that by 2008, 30% or more of mobile business in a variety of global markets depends on non-voice services. The evolution of packet/IP-based networks enables efficient development, control, integration, and delivery of IP multimedia services. At the same time, a converging service framework allows services to be created and delivered while providing access that is both open and secure.

Goals of future systems beyond 3G are straightforward—to provide wireless services to an increasingly mobile society that are dependable and enhanced, while minimizing their cost (per megabyte). Such systems will require higher speeds, higher performance, and higher capacity. There has been a flurry of activity to standardize, test, and implement next-generation systems beyond 3G. Each of these systems has relied on similar technology breakthroughs, which include advanced coding and modulation (such as adaptive space/time coding and 16 or 64 QAM), sophisticated antenna technologies (MIMO), high-capacity multiple-access mechanisms (OFDM), fast scheduling, and dynamic bandwidth and resource allocation. The notions of spectral efficiency, multiple smart beams, dynamic carrier structure, and dynamic resource allocation are all designed to provide much more capacity at lower cost.

Who is a Wireless Professional?

Each year, hundreds of schools in dozens of countries graduate thousands of wireless professionals. The education these institutions provide equips their graduates with varying levels of wireless system knowledge. Some provide basic and some provide advanced training, while others provide an in-depth education within a narrow specialty. Unfortunately, there is no common set of educational requirements that dictates the level of training.

Today, more than ever, the dynamic growth and globalization of the wireless communications industry brings to the forefront the need for all practitioners to rely on a common language and set of tools. The intent of the WEBOK is to serve as a tool to help develop common technical understanding, language, and approach among wireless professionals whose careers have developed in different parts of the world.

The Wireless Engineering Body of Knowledge

The WEBOK, produced by the IEEE Communications Society, is the product of a large international group of professionals, experts from both academia and industry.

The information presented in the following chapters is a general overview of the evolution of wireless technologies, their impact on the profession, and common professional best practices. Many wireless professionals may also find the WEBOK to be a useful tool for keeping pace with evolving standards. Appendix A includes a large number of references to books and articles that readers are encouraged to consult in order to enhance their knowledge and understanding of wireless technologies.

The WEBOK should not be viewed as a study guide for any wireless certification exam; it does not address all the topics that may be covered therein. Rather, it is intended as an outline of the technical areas with which a wireless practitioner, employed in industry, should be familiar, and offers suggestions as to where to turn for further information and study.

Organization

The WEBOK is organized into seven chapters:

- Chapter 1: Wireless Access Technologies

  Focuses on radio-access architectures and standards, and comments on the newest developments in wireless currently being used. It analyzes and compares many alternatives for radio access and classifies the different options according to the desired performance of the wireless solution.

- Chapter 2: Network and Services Architectures

  Focuses on the core network, supporting the access technologies described in the previous chapter. Concepts like switching, routing, and mobility management are among the chief topics covered.

- Chapter 3: Network Management and Wireless Security

  Summarizes common tools used to manage, control, and keep secure a wireless network. Concepts include service level agreements, configuration management, alarm handling, and providing security for a wireless network.

- Chapter 4: Propagation and Antennas

  Includes the central topics of radio frequency, engineering propagation, and budget calculations. Also presented are the architectures of many RF coding schemes, along with their relative advantages and disadvantages.

- Chapter 5: Facilities and Wireless Infrastructure

  Describes the common practices and the recognized international standards that should to be considered when designing a facility for active equipment.

- Chapter 6: Agreements, Standards, Policies and Regulations

  Reviews the policy mechanisms of the wireless industry that are necessary to anticipate, improve, and control the entities that design, implement, operate, and evolve wireless communication networks.

- Chapter 7: Wireless Engineering Fundamentals

  Lists the broad and basic technical knowledge that may be expected of a wireless practitioner.

The WEBOK is intended for a practicing wireless professional who has acquired at least the basic knowledge described in chapter 7, Wireless Engineering Fundamentals. If, on the other hand, a reader is aware of gaps in his or her skills and knowledge base, chapter 7 is an excellent way to begin addressing those deficiencies.

Chapter 1

Wireless Access Technologies

1.1 Introduction

Wireless links are broadly utilized in point to point, point to multi-point, and mesh applications, in fixed or mobile, satellite or terrestrial, as backhaul or as user access network, based on design goals, access technology definition, spectrum, and radio engineering principles. A great asset of wireless access is its enabling user mobility, whether at nomadic or at high speeds. Phased evolution of the user's true mobility is enabled through seamless connectivity at multiple levels. Geographically, the user may be connected through one or more personal (PAN), local (LAN), metropolitan (MAN – campus, hot-zone, municipality, mesh), or wide (WAN) area network(s).

There is further granularity in wireless access, increasingly enabled through sensing, mobile tags, and near-field communication ("touch" or scan zone). A user's mobility is maintained both through intra-technology and inter-technology handoff. The former may occur when moving from one cell to another in a cellular network, and the latter may occur when the user's session and application is maintained while the access moves from one technology (e.g. wireless LAN) to another (e.g., 3G). Although the user may have some level of awareness with respect to the access or connectivity mechanism, the user's communication space is ultimately (and increasingly) virtual, aware of intention, application, preferences, interaction, and experience, but generally not the access mechanism or network technology. This goal of creating an increasingly natural communication, which is more user-centric and less technology-centric, makes the enabling role of technologies and technologists more significant, more exciting, and perhaps more complex. In addition, a so-called natural communication may be enhanced by the application enabler, or the user terminal, as it discovers and utilizes smart system capabilities.

A wireless access network must obviously allow the end user(s) to access the network. This requires signaling, transmission, and communication aspects over wireless links, with coverage, capacity, and user experience defined by such attributes as session continuity, data rate, latency, security, and quality of service, among others. A group of users share system resources and are awarded access, governed by a certain discipline. At the heart of this discipline lies a multiple- access mechanism. How can an increasingly large number of users access the network, and the same channel, at the same time? Multiple-access mechanisms (e.g., FDMA, TDMA, CDMA, or OFDMA) have evolved through generations of wireless systems to enable this, with continuing improvement in data speed, capacity, and cost efficiency.

Coverage is particularly significant in wireless network design, in reach, indoor penetration, and continuity. Capacity is another design fundamental. This is an end-to-end attribute but greatly affected by the wireless access component. There is a need for small cell sites and more transmission carriers (and efficient use of bandwidth) to provide sufficient capacity for more users, or more accurately, greater simultaneous traffic. Generally, wide-area cellular networks are limited by coverage in low traffic areas, and by capacity in high traffic areas.

This brings us to the important notion of frequency spectrum. A radio tone has a frequency, and a radio signal carrying information has a range of frequency content. Modulation at the transmitter, based on and coupled with a given multiple-access mechanism, allows wireless communication over particular frequency bands. These bands are designated by local regulatory authorities, and generally coordinated by regional and global (International Telecommunication Union) bodies. They may be licensed (e.g., bands used by service providers in generations of mobile cellular technologies), or unlicensed (e.g., bands used by WLAN and Bluetooth, among others).

Mobile communications systems have traditionally been designed and optimized for voice communication. The first generation of wireless networks consisted of analog systems designed almost entirely for voice. Although voice continues to be the dominant application, data applications have grown dramatically over the years, from basic messaging, downloads, browsing, and positioning applications (enabled by second-generation (2G) systems and their enhanced versions) to an incredible growth of multimedia and content-based applications, particularly enabled by third-generation (3G) systems such as UMTS and CDMA2000. The core network (discussed in the next chapter) is evolving to provide a ubiquitous application environment with a common service architecture, and seamless access to multiple-access techniques. This is a phased evolution to an all IP or packet (heterogeneous) network with a flat (or flatter) architecture allowing seamless mobility across different access technologies.

ITU has defined the family of 3G systems (IMT-2000) and has set out the goals and attributes of the systems beyond 3G (IMT-Advanced). The standards bodies (e.g., 3GPP, 3GPP2, IEEE 802.x, and WiMAX Forum) have developed definitions for generations of mobile and nomadic communication access (and core) technologies, working with other standards groups such as IETF (to leverage Internet-related universal protocols and elements) and Open Mobile Alliance (to standardize service-enabler definition and interfaces).

This chapter starts with fundamental access network concepts and moves on to introduce access technologies and standards. As this is truly a broad topic, this chapter highlights key concepts and technologies but does not claim to be exhaustive, or inclusive of all forms of access technologies or implementations. Furthermore, it does not intend to promote or validate any particular technology. While significant technology attributes and design goals are highlighted, it must be noted that product innovations, implementation environment, customer solutions, operational ingenuity, and inter-carrier initiatives can provide a variety of prospects and capabilities above and beyond fundamental concepts, standards, technology goals, and common core practices.

1.2   Contents

This chapter addresses the following topics:

Design fundamentals
Mobility management
Definition of wireless access technologies
Digital mobile cellular technology evolution – 2G to 3G

1.3   Design Fundamentals

1.3.1   The Generic Picture

Wireless access technologies allow connectivity and communication over wireless link(s). They are based on principles of radio engineering: propagation, power, antenna technology, modulation, and link analysis.

Wireless access networks set up an intelligent wireless connectivity, with increasing sophistication in speed, performance, and efficiency, to enable user access, networking, and applications.

Figure 1-1 shows a generic wireless transmission system [1a] with the functions of transmission, propagation, and reception. The figure also shows an example of a wireless (mobile) access system architecture as a key component of an end-to-end communications network [1b].

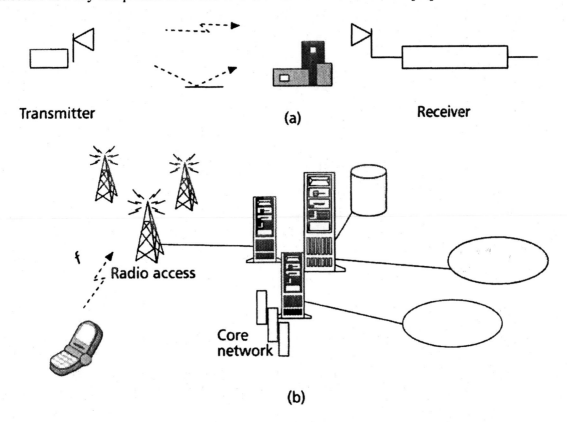

**Figure 1-1: Simplified View of a Generic Wireless Transmission System [a]
and a Cellular Network [b]**

Although oversimplified, this figure illustrates how all wireless systems (and their evolution) deal with transmission/reception (e.g., coding and modulation), antennas, link and propagation attributes, spectrum, and multiple-access etiquette. For example, an evolution from third- generation (3G) mobile-access technologies to beyond 3G is enabled by technologies that leverage advanced forms of coding and modulation, multiple-access dynamic resource allocation, spectrum management, and antenna technologies, among others.

.All wireless communication systems, satellite or terrestrial, fixed or mobile, personal, local, or wide-area, dedicated or shared, transport (backhaul) or access, regardless of frequency bands or topologies, have a similar fundamental anatomy. Furthermore, they have such similar concerns as coverage, capacity, transmitting power, interference, received signal power, infrastructure, and, of course, performance and efficiency. The detailed attributes, however, vary depending on design and application, access technology, links, mobility, or frequency spectrum. Details for the case of mobile cellular systems are provided in section 1.3.3.

1.3.2 Multiple-Access Mechanisms

FDMA

Frequency Division Multiple Access, or FDMA, is an access technology for sharing the radio spectrum. In an FDMA scheme, the radio-frequency (RF) bandwidth is divided into adjacent frequency segments. Each segment is assigned to a communication signal (e.g., related to each user in analog systems) that passes through a transmission environment with an acceptable level of interference from signals in adjacent frequency segments.

FDMA also supports demand assignment (e.g., in satellite communications) in addition to fixed assignment. Demand assignment allows all users apparently continuous access to the transponder bandwidth by temporarily assigning carrier frequencies on a statistical basis.

FDMA has been the multiple-access mechanism for analog systems. It is not an efficient system on its own, but can be used in conjunction with other (digital) multiple-access schemes. In this hybrid format, FDMA provides the frequency channel plan in which sophisticated digital multiple-access schemes are applied to each channel.

TDMA

Time Division Multiple Access (TDMA) is a channel-access method for shared-medium (radio) networks. It allows several users to share the same frequency channel by dividing the signal into different time slots. The users' information is transmitted in rapid succession, each individual using its own series of slots. This allows multiple stations to share the same transmission medium (the RF channel) while using only the part of its bandwidth that is required. TDMA is used in 2G digital-cellular systems. Examples include the Global System for Mobile Communications (GSM), IS-136, Personal Digital Cellular (PDC), iDEN, GPRS, and EDGE systems. It is also used in the Digital Enhanced Cordless Telecommunications (DECT) standard for portable phones and in some satellite systems. Figure 1-2 shows the TDMA mechanism.

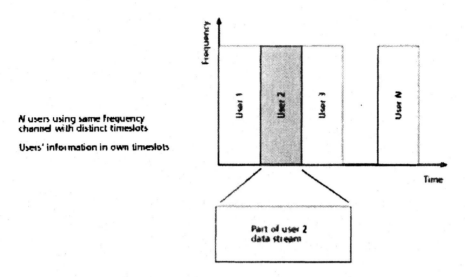

Figure 1-2: The TDMA Shared Mechanism

TDMA's features (and concerns) include simpler handoff and less stringent power control, while potentially allowing for more complexity in cell breathing (by borrowing resources from adjacent cells), synchronization overhead, and frequency/slot allocation (in comparison to CDMA). On its own, TDMA is limited by the number of time slots and the fast transition between them. However, in addition to its use in many existing systems, it has potential for being further leveraged in future hybrid multiple-access systems.

CDMA

Code Division Multiple Access (CDMA) is a channel-access method used by various radio-communication technologies. It employs a form of spread-spectrum and a special coding scheme (where each transmitter is assigned a code).The spreading ensures that the modulated coded signal has a much higher bandwidth than the individual user data being communicated. This in turn provides dynamic (trunking) efficiency, allowing capacity versus signal-to-noise ratio tradeoffs.

Multiple user signals share the same time, frequencies, and even space, but remain distinct as each is modulated (or correlated) with a distinct code. The codes are (quasi-) orthogonal such that a cross-correlation of a received signal with the "wrong" codes results in a spread (and hence suppressed) "noise," while the auto-correlation with the "right" code results in the (de-spread of the) desired output. The signal-to-noise power ratio decreases as the number of users increases, or the load on the system increases. This implies that with lower load, higher quality is achievable while, conversely, if some degradation is tolerable, the system allows higher capacity.

CDMA has been used in many communication and navigation terrestrial and satellite systems. Most notably, it has been used in third-generation (3G) mobile cellular systems (UMTS, cdma2000, TD-SCDMA) for its strong features such as capacity/throughput, spectral efficiency, and security, among others.

TD-CDMA and TD-SCDMA

Time-Division CDMA (TD-CDMA) and Time-Division-Synchronous CDMA (TD-SCDMA) use CDMA channels (5 MHz and 1.6 MHz, respectively) and apply TDMA by slicing the time.

TD-CDMA and TD-SCDMA are 3G technologies standardized by the 3rd Generation Partnership Project (3GPP) with different chip-rate options, UTRA TDD-HCR and UTRA TDD-LCR, respectively.

TD-SCDMA is being introduced in China. For more information, visit www.tdscdma-forum.org and www.tdscdma-alliance.org.

OFDM

Orthogonal Frequency Division Multiplexing (OFDM) is a multiplexing technique that subdivides the available bandwidth into multiple frequency subcarriers, as shown in Figure 1-3. In an OFDM system, the input data stream is divided into several parallel sub-streams of reduced data rate (and, thus, increased symbol duration) and each sub-stream is modulated and transmitted on a separate orthogonal subcarrier. The increased symbol duration improves the robustness of OFDM to delay spread. Furthermore, the introduction of the CP (Cyclic Prefix) can completely eliminate ISI (Inter-Symbol Interference) as long as the CP duration is longer than the channel delay spread. The CP is typically a repetition of the last samples of the data portion of the block that is appended to the beginning of the data payload.

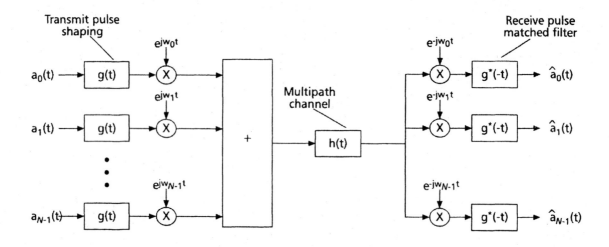

Figure 1-3: Basic Architecture of an OFDM System

OFDM exploits the frequency diversity of the multipath channel by coding and interleaving the information across the subcarriers before transmission. OFDM modulation can be realized with efficient IFFT (inverse fast Fourier transform), which enables a large number of low-complexity subcarriers. In an OFDM system, resources are analyzed in the time domain by means of OFDM symbols and in the frequency domain by means of subcarriers. The time and frequency resources can be organized into subchannels for allocation to individual users. The subchannelization can be referenced by the OFDMA mode of WiMAX standard 802.16E-2005.

Orthogonal Frequency Division Multiple Access (OFDMA) is a multiple-access/multiplexing scheme that provides multiplexing operation of data streams from multiple users onto the downlink subchannels, and uplink multiple accesses by means of uplink subchannels. This allows simultaneous low data rate transmission from several users. Based on feedback information about the channel conditions, adaptive user-to-subcarrier assignment can be achieved. If the assignment is done sufficiently fast, this further improves the OFDM robustness to fast fading and narrow-band co-channel interference, and makes it possible to achieve even better system spectral efficiency.

The OFDMA symbol may consist of a certain subcarrier structure as shown in Figure 1.4.

- Data subcarriers for data transmission
- Pilot subcarriers for estimation and synchronization purposes
- Null subcarriers for no transmission; used for guard bands and DC carriers

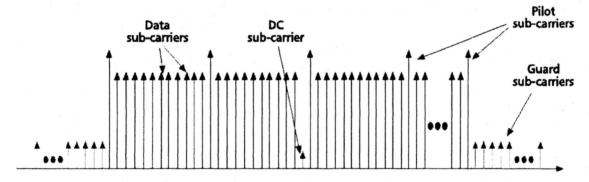

Figure 1-4: The OFDMA Subcarrier Structure

Active (data and pilot) subcarriers are grouped into subsets of subcarriers called subchannels.
OFDMA has certain elements of resemblance to CDMA, and even a combination of other schemes (considering how the resources are partitioned in the time-frequency space.

Put simply, OFDMA enhances the capacity of the system significantly and yet efficiently. Advanced OFDMA systems address such concerns as required flexibility in wide-area mobility, and complexity in adaptive subcarrier assignment, co-channel interference mitigation, and power consumption.

Advanced technologies beyond third-generation mobile take advantage of OFDMA's great potentials for significant capacity and efficiency improvement, together with other innovations (e.g., in coding and modulation and in antenna technologies).

1.3.3 Mobile Cellular Architecture & Design Fundamentals

Mobile cellular networks have grown rapidly since commercial services began in 1983. The network has evolved from pure circuit voice communications to high-quality voice and multimedia support and high-speed connectivity (access). The evolution of wireless mobile networks has been driven by the need to support mobile services, with evolving spectral efficiency and user experience.

A simplified wireless network architecture is illustrated in Figure 1.5. The user terminal is wirelessly connected to a Base Transceiver Station (BTS). This base station and a number of others are connected to a Base Station Controller (BSC). Traditional circuit voice is supported through a Mobile Switching Center (MSC) both directly (not shown) and in connection to a Public Switched Telephone Network (PSTN). The BSC can also be connected to an IP Gateway to support various packet data services. Quality of the wireless access connectivity is measured by call drop rate, access failure, block probability, packet loss rate and/or network reliability.

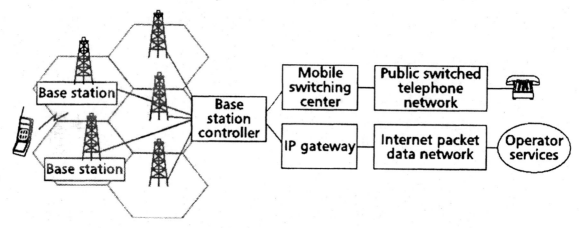

Figure 1-5: A Basic Wireless Cellular System

1.3.3.1 Capacity and Coverage Considerations

Capacity and coverage engineering are needed to optimize the system connectivity. Coverage is defined as the geographical area that can support continuous wireless access connectivity with the minimum guaranteed Quality of Service. It is heavily impacted by terrain, RF environment, and interference. Capacity is defined as the maximum number of users or the total data throughput a network can support reliably. It relies on traffic loading, traffic patterns, cell site equipment capability, and hardware dimensioning. The two crucial system attributes are typically interrelated.

Some access mechanisms have a theoretically deterministic capacity based on their channel structure (e.g., FDMA and TDMA systems), while others have dynamic channel allocation and allow some tradeoff of transmission quality vs. capacity (e.g., CDMA systems).

For example, a FDMA-based analog (AMPS) system has a channel bandwidth of 30 kHz. Therefore, a 10-MHz cellular band can support 333 FDMA channels. With a frequency reuse of 7, this is equivalent to a radio channel capacity of 15 channels per sector for a site with 3 sectorized cells. A TDMA system can further divide the time slot (typically) to 3 users and thus increase the capacity to 45 channels per sector.

CDMA capacity is a function of required signal bit-energy-to-noise-density ratio (E_b/N_o), spreading factor (chip rate B_{ss} divided by data rate R), channel activity factor (D), sectorization gain (G_s) and frequency *reuse factor* (K). The maximal number of users a CDMA sector can support, or the reverse link (uplink) pole capacity, is

$$N = 1 + \frac{B_{ss}}{R} \cdot \frac{1}{E_b/N_o} \cdot \frac{1}{D} \cdot G_s \cdot K \qquad (1)$$

For example, assuming 1.2288 Mb/s chip rate, 9.6 kb/s channel data rate, a frequency reuse factor of 0.66 and channel activity factor 0.4, the uplink (or reverse) channel capacity is around 36 traffic channels in a 3-sector cell for cdmaOne with 7 dB required E_b/N_o. This number increases to 72 for cdma2000-1x with the required E_b/N_o reduced to 4 dB. Typically, cdma2000 operational capacity is around 50% of those maximal pole capacity numbers due to forward link interference limitations. This means a commercial operational capacity of 36 in 1.25 MHz, or around 288 users over 10 MHz.

Coverage area is determined by the operating frequency, cell planning, radio receiver sensitivity, and required signal-to-noise ratio that an access technology can support. Typically, cellular network coverage is determined by the reverse link due to limited mobile station transmit power.

In a CDMA-based system, capacity and coverage are optimized by adjusting various power- management components. This includes sector-level and link-level power management. Therefore, a CDMA system must be optimized from a system point of view, so that the system can tolerate a maximal interference level. Figure 1.6 implies that when the system loading is above 75% of reverse-link pole capacity, as specified in Equation 1, the coverage will shrink dramatically.

This capacity and coverage tradeoff becomes even more important in support of IP multimedia services, where both affect the overall Quality of Service. Commercial cellular networks deployed worldwide have continuously grown through cell splitting and sectorization, in addition to technology advancements, to optimize capacity, coverage, quality, and cost considerations and trade-offs.

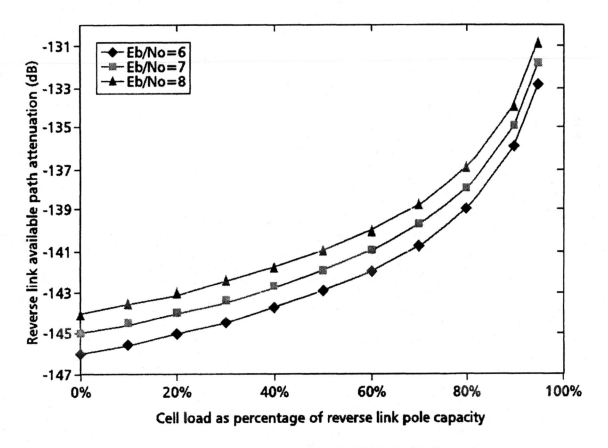

Figure 1-6: Capacity and Coverage of a CDMA Radio Network

It is important to note that as in any other engineering practice, innovation, and implementation, the desired service and experience should be targeted while maximizing the cost efficiency. As indicated earlier, generally, the design of a low-traffic area (cell) is governed by coverage, and that of a high-traffic area, by capacity, given performance requirements.

1.3.3.2 Spectrum Considerations

Wireless communication is obviously based on transmission of radio signals. The frequency spectrum is used for a variety of services, determined by propagation characteristics, and allocated to users according to the rules of the local spectrum regulatory authority. These rules may be influenced by a desire for competition, historical precedent, or the need to provide specific public services.

Some significant spectrum considerations, in relations to access technologies are briefly highlighted:

- Frequencies below ~4 GHz have proven suitable for non-line-of-sight (NLOS) applications such as mobile cellular. Higher frequencies however have also been used extensively (e.g., satellite communication).
- Higher frequencies have higher (free-space) power loss and shorter reach for the same transmit power, and tend to be more suitable for line-of-sight applications.
- Higher frequencies, where propagation is limited but more bandwidth is available, lend themselves to applications where capacity is a concern; lower frequencies such as 0.7–1 GHz may be used if coverage is the main consideration. In practice, a combination of considerations, especially spectrum availability, will determine what spectrum is used for an application.
- Lower frequencies (e.g., below 800 MHz) have high penetration properties, providing better indoor coverage.
- To meet growing capacity requirements (e.g., for mobile applications) sufficient frequency spectrum is needed. This is to meet broadband speed and high-traffic volume (in addition to technology) requirements. ITU-R leads in the identification of global spectrum needs for future systems.
- Uplink and downlink communication paths need to be divided (duplexed), either in frequency (FDD) or in time (TDD). Uplink and downlink frequencies are distinct and separated in the FDD case, as in most mobile cellular systems today. TDD systems, typically used in data systems, use the same bandwidth, with separate timeslots allocated for up- and down-link transmission.
- Wireless systems are designed with ingenuity particularly to avoid interference. Interference can potentially come from a variety of sources, including services operating in adjacent spectrum allocations. Use of guard bands between channels, systems, and at the FDD/TDD channel boundary is a standard practice.
- A significant consideration is regional and global frequency band alignment. This facilitates user roaming, and cost-effective standardized user terminal product availability.

As examples, mobile cellular operation in North America includes frequency bands at 850 MHz and 1900 MHz, while AWS (1700/2100 MHz) spectrum has recently also been auctioned. In addition, the 700-MHz spectrum has been auctioned in the U.S. The 2.5–2.7-GHz band is increasingly becoming a strong candidate band for broadband wireless systems. Currently the 2.4- GHz and 5-GHz bands are used worldwide, of which, Wi-Fi is the most widely known example. Differences exist, such as the broad use of 900, 1800, and 1900/2100 MHz in Europe and elsewhere vs. 850 and 1900 MHz in North America. In these cases subscriber terminals are designed to support multiple bands to support international roaming.

1.4   Mobility Management

1.4.1 Motivation

Mobility of the end user introduces certain important changes to the conventional telecommunication system. Since the user terminal is mobile, it is important to know where it is located at any given moment so that all outgoing and incoming calls can be handled efficiently. In a wireless network, this task is performed using a reference database known as the HLR (Home Location Register). The HLR not only contains information about the handset and its capabilities, but it also describes the base/home area of the user. Attached to the HLR is a VLR (Visitor Location Register) that provides information about the latest location of the user (or at which base station the handset is currently registered). This VLR is updated whenever the user moves from one area to another. The handset registers with the nearest cell sector allowing the system to update the VLR information. Both outgoing calls and incoming calls are thus handled through the base station to which the user is currently connected.

If the user travels while on a call, handoff occurs from one base station to the next, and the VLR is automatically updated at the end of the call. If the user travels *without* being on a call, the handset performs a reselection of base stations on the way. This reselection allows transfer registration (distance or zone based), which in turn helps update the VLR so that the system knows which base station must be contacted in order to send an incoming call. An incoming call is known as call termination and a paging message must be sent to the serving base station in order to get an initial response from the handset before sending a ringing alert indicating an incoming call. The following paragraphs describe the process of registration, paging, and other sequential procedures that the mobile/handset obeys that allow calls to be originated and terminated to the handset irrespective whether the user is stationary or mobile.

Mobility management encompasses a set of tasks for supervising and controlling the mobile user terminal (or mobile station, MS) in a wireless network. The tasks are divided into registration and paging, admission control, power control, and handoff (also called handover).

Registration is the process that informs the network of the presence and location of an MS. By paging, the network alerts the MS to an incoming call or message. Admission control determines when the MS gains access to the network based on the priority of the request compared to the availability of the network resources. Power control is necessary to keep interference levels at a minimum in the air interface, and to provide the required quality of service. Handoff handles the mobility when the user moves from the coverage of one cell site to another, or to the service of another wireless access technology.

Figure 1-7 shows the main states of a mobile station. Upon power up, the handset goes through an initialization state and acquires the preferred wireless network. The MS is in an idle state when not on an active call or connected to the network. The MS is in this idle state most of the time; it monitors overhead messages from the network or listens for incoming calls. System access refers to when the mobile accesses the network to set up a traffic channel for a voice or data call. The MS is in a connected or traffic state when it has a dedicated connection to the network for the transfer of voice or data packets.

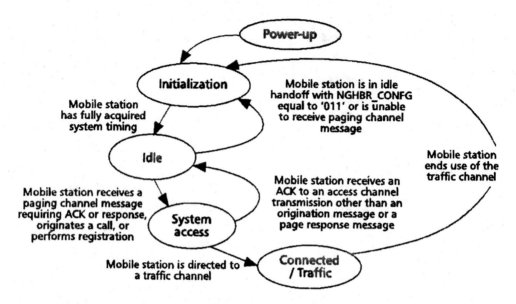

Figure 1-7: The Main States of a Mobile Station

1.4.2 Registration

Through registration, a MS notifies the cellular system of its location, status, identification, and capabilities. Registration also allows the network to efficiently page the MS when setting up a call.

There are two types of registration:

1) Autonomous: Triggered by an event or condition
   a. Power-up registration—MS powers on or connects to a serving system.
   b. Power-down registration— MS powers down, preventing unnecessary attempts by system to reach a user.
   c. Timer-based registration—MS registers when a timer set by the system expires. The system can de-register a MS that fails to register on power-down (because, for example, it is out of coverage range).
   d. Distance-based registration—MS registers when the distance between its current base station and the base station where it last registered exceeds a specified threshold. This is useful if the MS is not highly mobile.
   e. Zone-based registration—MS registers when it enters a new zone defined by the network operator. Some technologies allow the MS to maintain a list of zones in which it is registered.

2) Non-autonomous: Registration is explicitly requested by the base station or implied, based on other messages sent to the MS.
   a. Parameter change registration—MS registers when a specific parameter (e.g., the frequency band) has changed.
   b. Implicit registration—MS and base station exchange messages that convey sufficient information to identify the MS and its location.
   c. Ordered registration—Base station orders a MS to register (e.g., while on a traffic channel).

Not all registration methods are supported by a given network. It depends on how the vendor and operator have optimized the overhead signaling due to registrations. Registrations can place a heavy load on the reverse-access channels.

1.4.3 Paging

When a MS is powered on, it goes into an idle state following initialization and the network-selection process. In idle, it listens to the network for overhead messages containing network information or pages indicating that it is being called. The MS monitors what is referred to as the paging channel.

There is a link between the number of pages sent and the number of registrations. A balance is needed and the system is designed to efficiently page the MS for mobile-terminated calls. The more often the MS registers, the more precisely the network knows its location so that paging—finding out where the MS is located—can be targeted to a smaller group of cells, reducing messaging over the precious wireless channel.

1.4.4 Slotted Mode

Typically, a MS spends much of its time in idle mode. To conserve battery life and maximize standby time, wireless technologies implement a sleep or slotted mode of operation. In slotted mode, a MS powers down some of its electronics and periodically wakes up to check for new overhead or page messages indicating an incoming call. The network and MS must be synchronized so that pages for the MS are broadcast when it is awake. The longer the sleep period, the better the battery life. However, this may come at the expense of it taking longer to page the mobile to complete an incoming call. A sleep period of between 2 and 5 seconds is typically assigned.

1.4.5 Admission Control

The air link in a wireless system is a shared resource intended to support a finite amount of capacity and users. If new users could join the network indiscriminately, they would at some point start to have a negative impact on the wireless connections already established with other users. For example, in CDMA-based systems, as more users establish connections, loading increases on the forward and reverse links. In turn, the power levels for the existing connections must increase to overcome this loading increase. For connections at the cell edge, the mobile or base station may already be at maximum power so that with increased loading, the call quality cannot be guaranteed.

Admission control adds network intelligence to the call-establishment process. Thus, before adding a new connection or user, the system first ensures it has sufficient resources and will not affect existing customers. Because these resources can be on the forward or reverse links, both are considered separately and a call is admitted only if it passes both forward- and reverse-link admission controls. The attributes that can be part of admission control include:

- Noise rise (in reverse-link systems)
- BTS power (in forward-link CDMA systems)
- Codes (for CDMA systems)
- Available frequencies or time slots (FDMA or TDMA systems)
- Call-processing resources

To meet noise-rise criteria in the reverse-link, for example, new calls from a mobile would not be admitted by the admission-control algorithm if the resulting interference is predicted to be higher than a pre-defined threshold.

For voice calls, a rejection of a new user by admission control is treated as a blocked call. However, some technologies allow calls to be re-directed to a neighboring cell with overlapping coverage if it has the capacity. For data connections, it is possible for the network to downgrade the throughput of existing calls to allow more users access to the network.

1.4.6 Power Control

To achieve high capacity and quality, wireless systems must employ power control. The goal here is to minimize transmission power on both the forward and reverse links to conserve system resources, as well as to minimize interference to other users while meeting the minimum quality restraints.

In CDMA-based systems with all MSs assigned the same frequency, reverse-link power control is needed to ensure that each MS signal will be received at the cell site at the same level to deal with the well-known near-far problem. Because mobiles are always on the move, some are close to the base station while others are much further away. The close-in mobiles would send stronger signals back to the base station and cause interference on the reverse link unless their power is controlled. In addition, the system capacity would be maximized if the transmit power of each MS is controlled to be received at the base station with the minimum signal level required for keeping the system noise floor as low as possible. The forward link has a different problem. For similar performance, mobile stations near the cell edge need more power from the base station than those close to the base station.

Reverse-link power control is made up of an open loop, a fast closed loop, and an outer loop. Reverse-link loop power control is made up of both an open loop and a fast closed loop. Reverse-link open-loop power control is determined by the MS. The mobile measures the received power level from the base stations and adjusts its transmitter power accordingly. If it receives a strong signal, it decides that the path loss back to the base station is low and therefore lowers it's transmit power. The power required can be determined by a calibration constant that factors in cell loading, cell noise figure, antenna gain, and power-amplifier output.

Reverse-link closed or inner-loop power control is a function of the base station. The goal of the closed-loop component is for the cell to provide rapid corrections to the MS's open-loop estimate in order to maintain the optimum transmit power level. The cell measures the relative received power level of each of its associated MSs and compares it rapidly to an adjustable threshold. Each MS can then be instructed to increase or decrease its power. This closed loop corrects for any variation in the open-loop estimate to accommodate gain tolerances and unequal propagation losses between the forward and reverse links. For cdma2000-1x, closed-loop power control operates at 800 Hz on both the forward and reverse links; for WCDMA, it operates at 1500 Hz.

The reverse-link outer-loop power control maintains communications quality by setting the target for the closed loop by periodically adjusting a signal-to-interference (SIR) target, or setpoint, based on the frame error rate (FER). Low FER wastes capacity and the SIR target is decreased. A poor frame error rate affects quality and the SIR target is increased

Similar fast closed- and outer-loop algorithms are implemented on the forward link for cdma2000-1x and WCDMA. An example is shown in Figure 1-8.

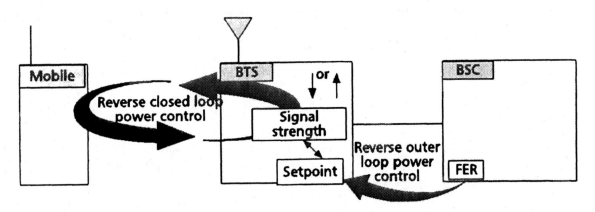

Figure 1-8: CDMA2000-1x Reverse Link Power Control

1.4.7 Handoff (Inter-technology & Intra-technology)

Handoff (or handover) is one of the key features with which a wireless network supports mobility. Handoff allows communications to be maintained between the MS and the network as the MS travels from the coverage area of one cell to another. As mentioned, the two primary states of a MS are the idle and connected modes and handoffs are necessary for both.

In idle mode, the MS monitors the network for changes in network information or for page messages for incoming calls. To receive these reliably, the MS should be communicating with the tower providing the best radio signal. This requires it to handoff as the MS moves and a stronger signal comes from a new tower. (Sometimes this process is referred to as cell re-selection.) The network broadcasts information in the overhead messages about the surrounding or neighboring cells. If it is in an idle state, the MS uses this information to periodically scan the strength and signal quality from a neighboring tower, and, if one tower signal is stronger, the mobile switches to that tower and then will continue to listen to the network. This handoff may trigger a registration criterion so that the MS will update the network with its new location. However, in many cases, the network will neither know nor be concerned that the MS is monitoring a new tower.

In connected mode, the MS has a voice or data connection with the network and reliable handoff is even more important to maintain quality service. There are two types of handoff: soft and hard. Soft handoff is also referred to as a "make-before-break" handoff and a hard handoff is known as a "break-before-make."

CDMA-based technologies use a soft handoff, which indicates that the MS is in communication with multiple base stations simultaneously, all with identical frequency assignments. This type of handoff provides diversity on both the forward and reverse links at the boundaries between base stations. While connected, the MS is continually searching for the presence of neighboring sectors or cell sites by monitoring the quality of their pilot channels. If a pilot of sufficient strength is detected, the MS will send a message with information about the new pilot to the communicating base station. If the base station opts for a soft handoff, it will check for and allocate resources at the new cell site and send a handoff message back to the MS directing it to perform a soft handoff. Similarly, if a pilot with which the MS is in soft handoff falls below a specific quality criterion, the MS will report this to the base station and the network will remove this connection to the MS. With soft handoff, as the MS moves from, say, cell A to B, cell B would be added to the MS as a handoff leg before cell A is removed; this is described as a make- before-break handoff mechanism. Softer handoff is a special form of soft handoff and applies when the MS is in handoff with multiple sectors from the same cell site.

Hard handoff occurs when the MS is moving between base stations that operate on different frequencies or with different technologies. FDMA- and TDMA-based systems utilize hard handoffs. In these situations, neighboring cell sites have different frequency assignments and the MS must retune itself to a new frequency before it can continue the connection. Hence, this is known as a break-before-make handoff. Hard handoff can also apply to CDMA-based technologies when multiple frequency assignments are present (e.g., when a second carrier is laid on top of the first for capacity needs).

There is a fundamental difference in handoff mechanisms between CDMA-based systems (cdma2000-1x or WCDMA) and TDMA systems (GSM). TDMA relies on discontinuous transmission, so there are gaps in time when the MS is not communicating with a base station. This offers an opportunity to make intersystem measurements with a single receiver on alternate frequency assignments or even with different technologies in the case of a GSM-to-UMTS handoff.

On the other hand, CDMA-based technologies rely on continuous transmission and reception. Therefore, WCDMA, as an example, introduces a compressed mode to create short gaps approximately a few milliseconds in both transmission and reception functions, and provides the mobile station an opportunity to make GSM measurements.

1.5   Wireless Access Technology Standardization

1.5.1 Motivation

Wireless access standards are designed to specify users' connectivity to networks and access to services through a user terminal, or a user interface to service client(s). In doing this, standards define the essential functions and protocols needed for access, coverage, throughput, and performance, as well as the mobility and roaming capabilities, interacting with the user terminal at one end and the core network, or back-end, at the other. An access technology is a transmission system that communicates through the wireless channels to achieve its connectivity and performance goals. It is effectively designed as a multi-user subnetwork with a multiple-access scheme, sharing of resources and fast scheduling of simultaneous users.

As expected, standardization aims to provide interoperability, inter-working, and, potentially, a rich set of attributes and a graceful evolution path. In addition, those who adopt a set of standardized technologies create an ecosystem that can provide prospects in availability of products and services, and economies of scale. It must be noted, however, that proprietary systems and their elements may also be introduced at different layers where there are opportunities for differentiation, ease of implementation, or time-to-market advantages.

The generations of wireless access technologies have increasingly been about standardized systems, building regional and global ecosystems, and enabling technology availability and user roaming.

1.5.2 Design Goals and Technology Elements

It is intuitively appealing to think of a wireless access technology standard as having certain design attributes such as:

- Connectivity, access
- Mobility, coverage, roaming
- Throughput, latency, performance, data symmetry
- Efficiency
- Universality (technology availability, global ecosystem, user roaming)

These attributes are fundamental but there are differences in how they are defined depending on the system's role and requirements:

- Low or high mobility
- Personal-area, local-area, hot-zone/metro, wide-area, near-field (scan-zone)
- Satellite or terrestrial; line-of-sight (LOS) or non-LOS
- Indoor coverage extension, or home network (e.g., femto cell)

So far, goals and requirements in this section have been identified from the viewpoint of a user or an application. Insight into the technology elements and functional capabilities should then address these goals and requirements. The evolution of technologies is typically about advancement of a set of technology elements and their end-to-end function. In particular, these technology elements include

- Coding, modulation, spectral efficiency, round-trip transmission, receiver structure
- Antenna technology, diversity, channel bandwidth and structure, FDD vs. TDD
- Multiple-access mechanism, user-access scheduling
- Resource sharing and allocation, power management, topology and distribution architecture

Although an access standard is generally not tied to a frequency band, frequency has a significant impact on the planning and development of an access technology. Choosing one band instead of another affect coverage (notably cell size), interference considerations, (indoor) signal penetration (e.g., at lower frequencies, below 900 MHz), line-of-sight constraints (e.g., mobile cellular frequencies below ~3 MHz), the user-terminal ecosystem and availability, and typically, capacity.

1.5.3 Technology Framework Definition and Standardization

The International Telecommunication Union (ITU) is the leading United Nations agency for information and communication technologies. It has three sectors, radiocommunication, telecommunication standardization, and telecommunication development, in addition to organizing global telecom events. The radiocommunication sector (ITU-R) in particular has several study groups on such topics as spectrum management, radiowave propagation, and satellite, terrestrial, broadcasting, and science services (http://www.itu.int/ITU-R).

A framework for the third generation (3G) has been defined in ITU's International Mobile Telecommunications–2000 (IMT-2000) family. The concept was born as far back as the mid- 1980's and the framework matured by 1999. This motivated global collaboration to define 3G standards with such design goals as flexibility, interoperability, affordability, compatibility, and modularity. The ITU-R M.1457 Recommendation identified five radio interfaces, while a sixth (WiMAX-based) air interface was added to the family in 2007:

- IMT-DS (Direct Sequence) – WCDMA/UTRA FDD
- IMT-MC (Multi-Carrier) – cdma2000
- IMT-TC (Time-Code) – UTRA TDD (TD-CDMA, TD-SCDMA)
- IMT-SC (Single Carrier) – EDGE
- IMT-FT (Frequency-Time) – DECT
- IMT-OFDMA TDD WMAN – WiMAX

The evolution of IMT-2000 beyond 3G (IMT-Advanced) is introduced later in this chapter.

The industry players in telecommunications, computing, broadcasting, user-terminal technologies, and applications have been involved for years in the development of wireless technology standards. With the growth of mobile communications, advances in technologies, availability of new spectrum, and the need for universal cost-effective solutions to enable rich systems and ecosystems, broad terminal availability, and the users' ability to roam, 3G standards have been defined through global partnership projects (PPs) involving standards bodies from different regions. Specifically, 3GPP and 3GPP2 have specified wireless access technologies (among others) for 3G and beyond.

The 3rd Generation Partnership Project, 3GPP, was established in 1998. The collaborating regional standards bodies, known as Organizational Partners, include ARIB, CCSA, ETSI, ATIS, TTA, and TTC. The project is run by the Project Coordination Group and its four Technical Specification Groups, each with a number of working groups (http://www.3gpp.org), as shown in Figure 1-9.

In its scope, 3GPP covers 3G systems based on advanced GSM core networks and the radio access technologies they support, namely Universal Terrestrial Radio Access (UTRA–both FDD and TDD). In addition, the partnership covers standards and reports for the maintenance and evolution of GSM technical specifications, and radio access technologies such as GPRS and EDGE. It has further defined systems beyond 3G and the phased evolution of IP networks. The work has been published in a number of 3GPP Releases (called R98, R99, Rel-4, Rel-5, etc.). The technologies are introduced in subsequent sections.

3GPP2 was born in parallel with and inspired by the 3GPP (and ETSI) efforts, and out of the IMT-2000 initiative. The partnership focuses on global specifications for systems (supported by ANSI/TIA/EIA-41) moving towards 3G (cdma2000) and beyond (UMB). Partners of 3GPP2 include the North American and Asian standards bodies ARIB, CCSA, TIA, TTA, and TTC. 3GPP2 has four Technical Specification Groups, each with several working groups, as shown in Figure 1-9. Standards developed by 3GPP2 are discussed in subsequent sections.

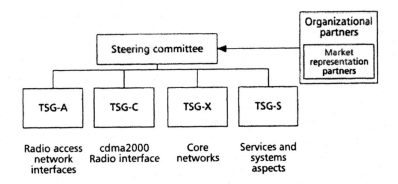

Figure 1-9: The Four Technical Specification Groups of 3GPP2

IEEE has been involved in creating a broad range of standards in a wide range of areas (http://standards.ieee.org). In particular, IEEE Project 802 (or 802 LAN/MAN) has developed local and metropolitan area network standards with a focus on the data link and physical layers (i.e., the bottom two layers in the OSI layered architecture) and the corresponding sub-layer structure. The many working groups within the family include ones that focus on wireless technology standards, extending to personal area networks (PAN):

- IEEE 802.11 – Wireless LAN
- IEEE 802.15 – Wireless PAN
- IEEE 802.16 – Broadband Wireless Access
- IEEE 802.20 – Mobile Broadband Wireless Access

These along with the working group for wireless sensor standardization form the so-called IEEE Wireless Standards Zone. Each work stream covers a range of specifications, typically evolving with versions and revisions. For example, the 802.16 working group has specified broadband wireless access across all scenarios of fixed, nomadic, and high user mobility. Furthermore, industry forums work extensively to define, certify, and promote related technologies, a variety of profiles (i.e., candidate parameter sets) and products. The WiMAX Forum, with a number of working groups, certifies and promotes the compatibility and interoperability of broadband wireless products based on the harmonized IEEE 802.16/ETSI HiperMAN standard (http://www.wimaxforum.org).

1.5.4 Mobile Technology Generations and Nomadic Implementations

Figure 1-10 shows a simplified view of the evolution of mobile access technology standards, and highlights nomadic broadband wireless access. Not all the technologies and evolutionary steps are shown. Moreover, given the fluid nature of innovations and implementations, the generation label may not necessarily represent each scenario accurately.

To help understand these technologies, the GSM evolution path, IS-95 (cdmaOne) roadmap, IEEE 802.11 (WLAN), IEEE 802.16 (WiMAX), and OFDMA-based technologies beyond 3G are outlined in some detail in subsequent sections. Also presented are introductions to personal, home, and near-field communications.

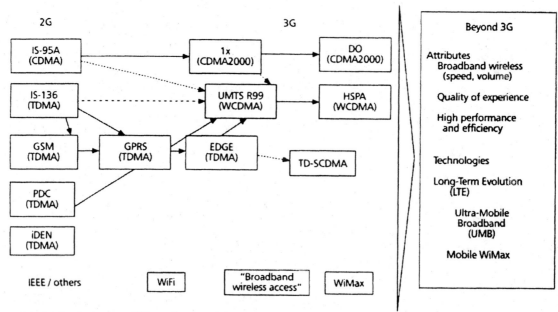

Figure 1-10: Evolution of Mobile Access Technologies into the Third Generation and Beyond

1.6   Digital Mobile Cellular Technology Evolution–2G to 3G

This section discusses in some detail a number of global wide-area mobile access technologies, including 2G through 3G systems based on TDMA and CDMA. OFDMA-based wide-area technologies are discussed later.

Again, note that the goal of this chapter is to provide an overview of wireless access, with some detail on terrestrial and global mobile and nomadic systems. The chapter neither includes all technologies and scenarios nor provides equal detail on all technologies. Furthermore, each implementation has its own attributes and reasons based on its market, context and history, goals, and roadmap. Actual parameters may also vary from those represented in standards or presented here.

1.6.1 3GPP Wireless Access Technologies

The evolution of GSM and the wireless access standards specified by 3GPP are divided here into three subsections covering the widely implemented TDMA-based technologies of GSM, GPRS, and (briefly) EDGE, followed by UMTS Phase 1 and HSPA. As indicated earlier, movement between different paths is fluid and based on needs and roadmaps. For example, operators with PDC or 3GPP2 technologies may have chosen to implement UMTS technologies or a 3G operator may choose any of the OFDMA-based technologies in its roadmap.

1.6.1.1 GSM, GPRS, EDGE

GSM/UMTS standards and networks range from GSM phase 1 to the HSPA family. GSM Phase 1 is a circuit-switched mobile network technology using TDMA, which provides voice services and short-message service (SMS). Subsequent phases of GSM introduced packet services (GPRS) while keeping such fundamental features as TDMA radio transmission, the MAP signaling protocol for roaming, and the security features. UMTS phase 1 (often referred to as Release 99) has kept the network principles of GSM and GPRS but has a completely new radio access interface based on CDMA.

Main principles of GSM

GSM is a mobile digital technology developed in several phases. Although it is a 2G system, the main principles of GSM phase 1 were set as early as 1987. It was optimized primarily to provide circuit-switched voice services, though basic data services, notably SMS, were soon introduced.

GSM radio interface

The GSM radio interface is based on the FDD mode and TDMA with 8 time slots per radio carrier. In other words, each uplink time slot is paired with a downlink time slot. Each radio carrier requires a 200-kHz uplink and a 200-kHz downlink. A time slot may be used for one of the following set of logical GSM channels, although there are other possibilities: 1 TCH (traffic channel) full rate, 2 TCHs half rate, 8 SDCCHs (stand-alone dedicated control channels), or 1 CCCH/BCCH (common control channel/broadcast control channel).

The TCHs are used to transmit voice or data while the SDCCHs can only be used for signaling or SMS transmission.

In Europe, the Middle East, Africa, and Asia, the GSM system generally operates in the following bands:

900 MHz (174 radio carriers): Uplink: 880-915 MHz   Downlink: 925-960 MHz
1800 MHz (374 radio carriers): Uplink: 1710-1785 MHz   Downlink: 1805-1880 MHz

In the Americas:

1900 MHz (298 radio carriers): Uplink: 1850-1910 MHz   Downlink: 1805-1880 MHz

Note that GSM is defined independently of the frequency resources, which means that it operates in certain other bands as well, notably at 850 MHz. (e.g. North America).

A generic example illustrates what this means for a GSM operator, whose typical spectrum allocation may be 20 MHz, sufficient for 100 carriers. A frequency reuse scheme is generally used with the result that around 8 carriers may be available in each cell. This means there are 64 TDMA time slots per cell of which 4 slots may be used for signaling traffic (for instance, 1 time slot for the CCCH/BCCH and 3 others to provide 24 SDCCHs). Therefore, 60 time slots are available in each cell for voice traffic. Using only TCH/FS allows 60 simultaneous voice calls. However, if TCH/HS alone are used, 120 simultaneous voice calls are possible.

Protocol Aspects
The signaling protocols of the radio interface are divided into a three-layer structure. This is similar to that of the DSS1 (digital subscriber signaling system 1 protocols on the D channel in ISDN, which is based on the OSI reference model.

According to the configuration requested by the mobile station, layer 2 offers either connectionless information transfer in unacknowledged mode (on point-to-multipoint or multipoint-to-point channels) or connection-oriented information transfer in acknowledged mode on a dedicated control channel. Layer 3 of the radio interface is divided into three sub-layers: radio resource (RR), mobility management (MM) and connection management (CM). The CM sub-layer comprises parallel entities: supplementary services handling, short message services, and call control (CC).

GSM Phase 1 Architecture
A cell's radio coverage is provided by a base transceiver station (BTS). Each BTS is linked to a BSC (base station controller); a BSC and the BTSs linked to it constitute a BSS (base station subsystem). Each BSC is linked to a MSC (mobile service switching center), as shown in Figure 1-11. The interface between the BSS and the NSS (network subsystem) is called the A interface. Seen physically, information flows between mobiles and BTSs, but seen logically, the mobile communicates with entities in the BSS and MSC. Layer 1 and layer 2 are handled by the BTS, the RR sub-layer is handled by the BSC and the MM and CC sub-layers are handled by the MSC.

There are in fact slight exceptions to these general rules but this mapping essentially means:

- The BTS handles the radio transmission.
- The BSC organizes the allocation, release, and supervision of the radio channels according to commands received from the MSC.
- The MSC handles call establishment and call release, the mobility functions, and everything related to the subscriber's identity.

Schematically, the behavior is as follows: the MS makes a first access on the RACH (random access channel, a multipoint-to-point channel) of the selected cell. In response, the BSC allocates the MS a first dedicated channel. After the MS has seized this radio channel, a dedicated link exists between the MS and the BSC. The MS uses this link to send an initial message, which includes its identity and the reason for the access (e.g., a requested service). Upon receiving this message, the BSC establishes a signaling connection with the MSC dedicated to this MS.

Key functional elements of GSM's core network include the mobile switching center (MSC), the home location register (HLR), and the visited location register (VLR). The MSC has an interface with the BSC, on the access side, in addition to the back-end and fixed networks, as indicated earlier. The MSC is a digital exchange, able to perform all functions for handling calls to and from mobile subscribers in its area, and to cope with the mobility of the subscribers using HLR and VLR. The signaling exchanges between these entities are specified in the mobile application part (MAP), which is the protocol used to provide roaming.

Core network technology and networking mechanisms are subjects of the next section.

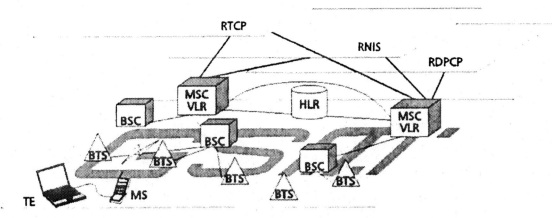

Figure 1-11: Basic Cellular (GSM) Network Architecture

SIM Features in GSM

The GSM mobile station, or user terminal, has two distinct elements: the mobile equipment that connects with the network and a chip card (the subscriber identity module, or SIM) that holds all subscriber-related data, and is used in obtaining access to the network A standard interface sits between these two elements. Data stored in the SIM include:

- The international mobile subscriber identity (IMSI).
- The Ki key, which is linked to the IMSI; it is allocated at subscription and stored unchanged.
- Algorithm A3 and algorithm A8.

These data are used for two security features—the authentication procedure and ciphering. Authentication enables the network to validate a mobile subscriber's identity, and protects the network against unauthorized use. When the MSC receives a mobile identity (IMSI) transmitted on the radio path, it triggers an authentication procedure. The network sends a random number RAND to the mobile station to check that it contains the Ki linked to the claimed IMSI. The mobile station applies algorithm A3 to RAND and Ki in order to compute the answer to be sent to the network.

The ciphering procedure prevents an intruder from listening to what is transmitted over the radio interface. This protection covers both the signaling and the user data for voice and non-voice services. The layer 1 data flow transmitted on dedicated channels (SDCCH or TCH) is the result of a bit–per-bit addition of the user data flow and of a ciphering stream generated by the ciphering/deciphering algorithm. This algorithm uses both a ciphering key and the TDMA frame number. The ciphering key (Kc) is computed independently on both the MS side and the network side by the authentication procedure; algorithm A8 is used to derive Kc.

The SIM card is offered by the network operator so the operator can provide the security features even during roaming. In addition, the SIM serves as a tool to support other features and services.

GPRS and EDGE-GSM Evolution

After the success of GSM phase 1, new features had to be defined in GSM networks. These included service consistency when roaming outside the home network, enhanced throughput to support data services and the ability of the SIM to become an active device able to control the MS (a SIM toolkit).

These new features were developed to be compatible with legacy user devices, and their introduction was optional.

Introduction of Packet Mode in GSM

Data services were already defined in the first phase of GSM. These were circuit-switched data services that yielded the allocation of a TCH (one TDMA time slot). Throughput was very low, typically 9.6 kb/s. Therefore, it was highly desirable to increase the throughput on the radio interface. One way to achieve this for a given call is to utilize more than one TDMA time slot on the radio interface. Such a feature has been standardized as HSCSD (high-speed circuit-switched data). The main advantage of HSCSD is its simplicity. The GSM MSC and A interface are unchanged as long as four or fewer time slots are used because in such cases a 64-kb/s circuit is sufficient on the A interface. The drawback of the HSCSD is that it uses radio resources inefficiently.

To address the concern for efficiency, another service was designed—GPRS (General Packet Radio Services). The GPRS is a set of GSM bearer services that provides packet-mode transmission and interworking with external packet data networks. The GPRS allows a subscriber to send and receive data in an end-to-end packet-transfer mode without utilizing network resources in circuit-switched mode. The service aspects of GPRS are specified in GSM 02.60 and its technical realization is specified in GSM 03.60.

To accommodate sporadic transfers of large amounts of data, GPRS encompasses allocation and release mechanisms that optimize use of the radio resources. During a data call, these resources (i.e., one or more time slots) are allocated only when data must be transmitted. Afterward, they are released although the data-transfer session can be kept going between the mobile station and the network. GPRS is well suited to asymmetric data transfer. To consider those, GPRS can allocate more TSs in the DL than in the UL.

GPRS was designed to allow a smooth sharing of the radio resources between speech and data. To accommodate data traffic, the operator can decide for each radio carrier how many time slots to allocate to GPRS traffic and how many to voice traffic. Sharing can be performed dynamically.

With a time slot, the following throughputs can be obtained on the radio interface according to the different coding schemes:

CS-1: up to 9.05 kb/s   CS-2: up to 13.4 kb/s   CS-3: up to 15.6 kb/s   CS-4: up to 21.4 kb/s

More than one time slot can be allocated for a data transfer. Typically, a GPRS mobile station may allocate up to 4 TSs in the downlink and 2 TSs in the uplink, which adds up to 85.6 kb/s in the downlink.

GPRS Architecture

GPRS was also designed to allow operators to reuse the radio coverage deployed for voice. In addition, it was necessary to define two new functional entities on the core network side:

- The serving GPRS support node (SGSN) is the node controlling the BSC and serving the MS. It handles mobility, paging, and security and interfaces with the BSC.
- The gateway GPRS support node (GGSN) works with the packet data networks (fixed and mobile). Connected to the SGSN through an IP network, it contains routing information for GPRS users.

These network entities have been defined to work with IP networks.

A Network with Packet-Switched and Circuit-Switched Domains
After the rollout of GPRS, a GSM network includes circuit-switched and packet-switched domains. The basic architecture is shown in Figure 1-12. (A detailed discussion of networks and networking are given in the next section.)

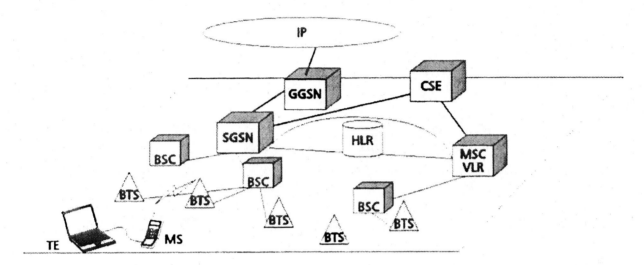

Figure 1-12: An Evolved GSM Architecture with Introduction of GPRS

EDGE
EDGE (enhanced data rates for GSM evolution) uses advanced modulation to increase the throughput at the radio interface. With a time slot, and depending on the different schemes, three throughputs can be obtained on the radio interface: 28.8 kb/s, 32.0 kb/s and 43.2 kb/s. As for GPRS, a MS can be allocated more than one time slot.

Using EDGE time slots with the GPRS architecture provides packet services with increased data throughput, also known as enhanced GPRS.

Further advancements are being developed to enhance EDGE capabilities (EDGE[+]).

1.6.1.2 UMTS Phase 1

The UMTS (Universal Mobile Telecommunication System) was designed as a 3G system in a joint effort of the GSM community. It is compatible with GSM to a considerable extent and meets the goals of IMT-2000 set by the ITU. This common work was organized in a joint standardization forum known as a Partnership Project (in this case, 3GPP).

The UMTS phase 1 specifications are based on the following key principles:

- A completely new radio interface using wideband CDMA (WCDMA).
- Reuse of the network services and security principles of GSM and GPRS.

Radio Interface

Whereas GSM radio technology is based on TDMA, the design of the UMTS radio interface is based on CDMA technology. A completely new radio interface was designed by ETSI and developed within 3GPP. This interface is formed by two modes of operation in two different parts of the spectrum: the FDD in paired-band configurations (for uplink and downlink) and the TDD in an unpaired one.

The FDD component is built on a WCDMA (wideband CDMA) concept based on direct-sequence CDMA with a 3.84 Mc/s chip rate, and is designed for flexible 3G services and optimized GSM compatibility. The physical layer offers flexible multi-rate transmission and a service-multiplexing scheme. Efficient support for packet access is defined with a dual-mode packet-transmission scheme supporting various multimedia services.

The TDD component is a TD-CDMA scheme, with both time and code multiplexing. It uses joint detection in the receiver on the uplink as well as downlink, and requires neither high-power control accuracy nor a soft handoff. This mode offers flexible downlink and uplink time-slot allocation to meet asymmetric traffic requirements and simplify the implementation of adaptive antennas. Basic system parameters such as carrier spacing, chip rate, and frame length are harmonized. Thus, FDD/TDD dual-mode operation is facilitated, which provides the basis for the development of low-cost terminals.

Each UMTS WCDMA FDD radio carrier requires a 5-MHz DL and a 5-MHz UL. A UMTS WCDMA FDD radio carrier allows around 50 simultaneous voice calls in circuit mode. Nevertheless, as mentioned earlier, the CDMA scheme provides trunking efficiency with no fixed time or frequency slots per user. This allows tradeoffs between traffic load and performance attributes, such as the throughput and the level of tolerable degradation. UMTS specifications provide the details. In Europe, the Middle East, Africa, and Asia UMTS generally operate in the FDD 1920–1980 MHz (UL) and 2110–2170 MHz (DL) bands, though other bands are possible (e.g., re-farming of 900 MHz). In the Americas, different frequency bands are used, notably around 850 MHz and 1900 MHz.

The split of the radio interface protocol stack into three layers is similar to GSM. Layer 1 is the physical layer with the new radio technology (WCDMA and TD-CDMA). Layer 2 is split in two sub-layers: the MAC sub-layer and the RLC sub-layer. Layer 3 encompasses three sub-layers: RRC (radio resource control), MM (mobility management), and CC (call control). The FDD component of the WCDMA radio interface provides radio bearers both for circuit-switched (e.g., voice) and packet-switched communication, up to 384 kb/s (DL) and 128 kb/s (UL) in UMTS Phase 1. The radio interface protocol architecture is specified in UMTS 25.301.

UMTS Phase 1 Architecture

The architecture of the UMTS radio-access network (UTRAN) is fundamentally similar to the GSM architecture in terms of node B (base station) and radio-network controller (RNC) components. The functional split is, however, somewhat different from that of the GSM architecture.

The core network of UMTS Phase 1 applies the principles of GSM and GPRS. It is comprised of MSC, VLR, HLR, SGSN and GGSN (introduced earlier) with a functional split similar to GSM's. As in GSM, messages are exchanged directly between the mobile terminal and the core network without any translation by RNC or node B. All UMTS radio-resource control functions are located in the RNC, including handoff functions. This is different from GSM, where control of the handoff process is shared between the visited MSC and the BSC. The circuit domain is quite similar to GSM's, although there has been evolution in the packet domain. Of particular note is that the GPRS interface between SGSN and BSC (Gb) is completely changed. Indeed, a tunnel is established between SGSN and RNC. Unlike GSM, the transcoders are controlled by the MSC and not by the access sub-network. Lastly, the SIM is kept in UMTS and is known as USIM (UMTS SIM).

Specifications UMTS 23.101 and UMTS 23.121 give general UMTS architecture principles and architecture requirements for UMTS Phase 1 (Release 99).

GSM/UMTS Interworking

GSM/UMTS handoff and roaming were defined from the start. The protocol for roaming (MAP) is common to the two. In addition, UMTS mobile stations typically support GSM at a minimum, and generally support multiple modes and multiple bands. A core network was also developed that's common to GSM and UMTS, allowing the GSM/UMTS operators to deploy a single core network enabling the different access technologies, as shown in Figure 1-13.

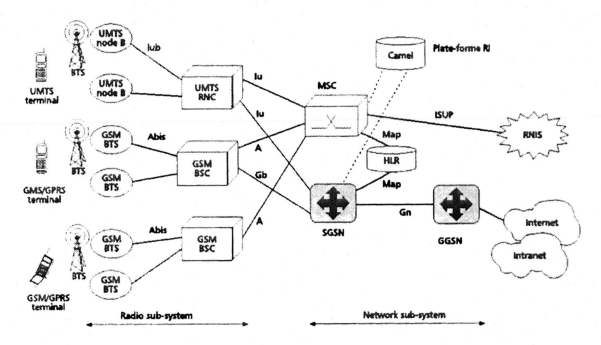

Figure 1-13: GSM Radio Access and UMTS Radio Access with the Same Core Network

1.6.1.3 HSPA (UMTS Evolution)

UMTS Release 5 defines the evolution of the UMTS access network, HSDPA, with its significant enhancements to downlink throughput capabilities. To improve the data throughput in the uplink, HSUPA was defined in UMTS Release 6. The different phases are generally defined with much consideration to the smooth and graceful evolution of phased implementations.

HSDPA (High-Speed Downlink Packet Access)
The growing importance of IP-based applications for mobile access made it necessary to develop the UMTS radio interface further to meet the requirements of new and expected applications:

- Packet data transmission
- High-throughput with asymmetry between uplink and downlink
- Low time constraint and non-uniform quality of service requirements (traffic in bursts)

This led to the standardization of HSDPA by 3GPP, in its Release 5, with two major improvements:

- The optimization of the spectral efficiency by a better allocation of radio resources, which enables a variety of speed profiles depending on the number of codes, modulation, and channel coding, and could reach a theoretical peak data rate of 14.4 Mb/s.
- The reduction of transmission delays.

HSDPA is an optimization of UMTS. The same frequency bands used for WCDMA Phase 1 may be used.

A fundamental goal of HSDPA, as is evident from its name, is to provide high throughput in the downlink. To achieve this it uses methods known from GSM/EDGE standards, including link adaptation and the hybrid ARQ (automatic repeat request) retransmission algorithm. In addition, HSDPA exploits a new dimension not exploited by other 3GPP systems: multi-user diversity that takes advantage of the multiplicity of users to optimize the use of radio resources. This multi-user diversity is extracted using a fast-scheduling algorithm, built in Node B. This scheduler selects at very short intervals the most appropriate MS (or user equipment) to which data should be sent.

Radio Interface
A simplified view of HSDPA and its basic functions is in Figure 1-14. Node B estimates the channel quality of each HSDPA user based on, for instance, power control, ACK/NACK ratio (i.e., the ratio of packet transmission acknowledgements to non-acknowledgements) and HSDPA-specific user feedback (CQI, the channel quality indicator). Scheduling and link adaptation are then conducted at a fast pace that depends on the scheduling algorithm and the user prioritization scheme.

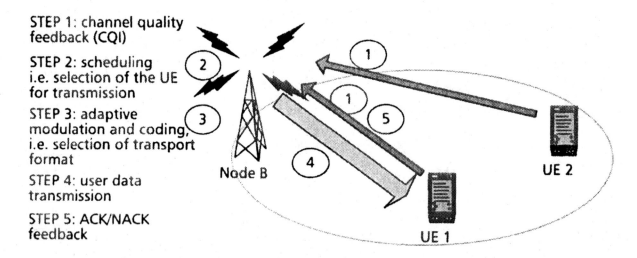

STEP 1: channel quality feedback (CQI)

STEP 2: scheduling i.e. selection of the UE for transmission

STEP 3: adaptive modulation and coding, i.e. selection of transport format

STEP 4: user data transmission

STEP 5: ACK/NACK feedback

Node B

UE 1

UE 2

Figure 1-14: Basic Functionality of HSDPA Access

HSDPA relies on a new downlink transport channel, the high-speed downlink shared channel (HS-DSCH). This channel is supported by new physical channels with distinct functions for data traffic or signaling on the uplink or downlink. Details are provided in 3GPP specifications.

In HSDPA, two fundamental features of UMTS/WCDMA, variable SF (spreading factor) and fast power control, are disabled and replaced by adaptive modulation and coding (AMC), extensive multi-code operation and a fast and spectrally efficient retransmission strategy. This allows selecting, for users in good radio conditions, a coding and modulation combination that provides better throughput with the same transmitted power as for Release 99 channels. To enable a large dynamic range for the HSDPA link adaptation and to maintain good spectral efficiency, a user may simultaneously use up to 15 codes of SF 16. In addition, a new modulation is introduced, the 16-QAM, in addition to QPSK. The use of more robust coding such as fast hybrid automatic repeat request (HARQ), and multi-code operation remove the need for variable SF. Table 1-1 shows some possible modulation and coding schemes.

CQI value	Transport block size	Number of codes	Modulation
1	137	1	QPSK
2	173	1	QPSK
7	650	2	QPSK
10	1262	3	QPSK
15	3319	5	QPSK
16	3565	5	16-QAM
30	25558	15	16-QAM

Table 1-1: Modulation and Coding Schemes (MCS) for Different CQI Values

To allow the system to benefit from short-term radio channel variations, packet-scheduling decisions are done in Node B. If desired, most of the cell capacity may be allocated for a very short time to one user, when radio conditions are the most favorable. Thus, an MCS providing high user payload (i.e., a larger transport block size) can be used. The classical 10-ms frame of UMTS has been divided in 5 sub-frames, or TTIs (transmission time intervals) of 2 ms.

In the optimum scenario, *scheduling* tracks fast-fading of user equipment (UE), as shown in Figure 1-15.

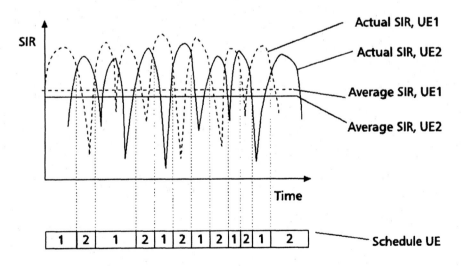

Figure 1-15: Example of Scheduling the User with Maximum Signal-To-Interference Ratio (SIR)

Choosing the appropriate scheduling algorithm is crucial for deploying HSDPA efficiently, since this algorithm has a direct effect on the throughput achieved by end users. Several types of algorithms may be implemented, from fair resource sharing to efficient algorithms that aim at maximizing cell throughput. A compromise is to use algorithms that try for a middle ground between fairness and efficiency, albeit at the expense of increased complexity.

The physical layer packet combining, HARQ means that the terminal stores the received data packets in soft memory. Then if decoding has failed, the new transmission is combined with the old one before channel decoding. The retransmission can be either identical to the first transmission (chase combining) or contain different bits compared with the channel encoder output received during the last transmission (incremental redundancy). With this incremental redundancy strategy, it is possible to achieve both a diversity gain and an improved decoding efficiency.

Radio Access Network Architecture

In 3GPP Release 5, the architecture of the radio-access network with HSDPA remains the same as in Release 99, with Node B and RNC. However, some upgrades were needed in nodes and interfaces to deploy HSDPA compared to UMTS R99. The major impact on access network nodes is that the HS-DSCH is terminated in Node B instead of in the RNC. This means that a small part of the 3GPP R99/4 functions, initially located in the RNC, has been moved down to Node B: e.g., the H-ARQ function (as part of the MAC layer) and the scheduling function, which are implemented in a new MAC entity. Therefore, software upgrades are necessary in Node B and RNC. Concerning interfaces, the major issue is one of dimensioning: since HSDPA provides higher bit rates (Iu-PS and Iub), interfaces should be re-dimensioned to support a higher throughput than for UMTS R99.

Concerning the MS (or UE), enhancements were necessary. For example, the new HARQ functions require more buffering and computational power; moreover, advanced receivers are needed, as the traditional rake receiver cannot decode more than five simultaneous codes efficiently.

A significant trend has been to move toward a flatter architecture to enable seamless mobility across different access networks. In 3GPP Rel7 and Rel8, enhancements to the architecture are introduced that enable deployment of a so-called flat architecture by collapsing Node B and RNC into a single node.

Physical Layer Performance

Cell and user throughputs depend on several factors:

- Radio link conditions: A user with good radio link quality will be allocated a modulation and coding scheme (MCS) that will provide a higher throughput than if conditions were poor.
- Scheduling algorithm: This has a major impact, as discussed earlier.
- Load on the cell: The user is scheduled more frequently and throughput is higher with less cell load.
- UE category: 12 categories of HSDPA-capable UEs have been defined by 3GPP, according to their capacity to simultaneously decode several codes, the modulation they can support, and the

periodicity at which data can be scheduled. The maximum achievable bit rate for a UE varies depending on these parameters, from 1.8 Mb/s to a theoretical peak of 14.4 Mb/s.

- Power allocated to HS-DSCH: The power allocated to HSDPA channels may be set to a fixed value or equal the power available after power is allocated for R99 dedicated and control channels.
- Number of codes allocated to HS-DSCH: Depends on UE category but also on the operator's configuration. For example, fewer codes are available if a carrier is shared with R99 DCH.

HSUPA (Enhanced DCH)

HSUPA is actually identified in the 3GPP specifications as enhanced DCH (E-DCH) or enhanced uplink. It is a feature added to UMTS in 3GPP Release 6 to handle high-data-rate packet services in the uplink, with a peak (target) data rate of 5.8 Mb/s.

The motivation for developing HSUPA was fundamentally the same as for HSDPA, but the need for enhanced uplink speeds was considered. HSUPA handles the radio and network resources better, as well as the transmission time, in the uplink.

Unlike HSDPA, HSUPA is not based on a shared channel but rather on an optimization of the R99 uplink DCH with traditional DCH features:

- A new enhanced dedicated transport channel, E-DCH.
- Power control to adapt E-DCH to a changing environment.
- The transport format selected according to current buffer status and available power.
- A soft handoff that benefits from inherent uplink macro-diversity.
- Predictable bit rate.

Moreover, some HSDPA-like enhancements are introduced:

- MAC functions are moved from the RNC to Node B to reduce the round trip time (RTT).
- Short uplink TTI is introduced (2 ms as an option, 10 ms as default), which also reduces RTT.
- Improved re-transmission (H-ARQ) is used to benefit from combining the received packet versions.
- Multi-code operation and low spreading factor are supported to increase the peak data rate, with the definition of several UE categories corresponding to various multi-code and SF configurations.
- Node B scheduling is applied to control the uplink interference level, the cell capacity and the UE QoS:
  - Node B dynamically allocates scheduling grants to *all* users in the cell, according to their rate requests, QoS requirements, uplink cell load, and maximum uplink target cell load.

The process of scheduling is illustrated in Figure 1-16. It is, by nature, less efficient than in HSDPA because the entity doing the scheduling (Node B) is not the one that allocates the resources (the UE).

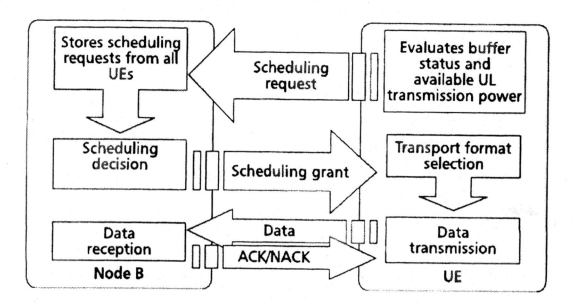

Figure 1-16: A Simplified View of HSUPA Node B Uplink Scheduling

One UE has at most one E-DCH transport channel that can be mapped over a variable number of physical channels. Other physical channels in the uplink and downlink are for traffic and signaling control.

Macro-diversity

As in R99 and contrary to HSDPA, soft and softer handoffs are supported for the E-DCH channel. This feature exploits the available uplink macro-diversity to improve the radio-link transmission for users at the cell edge, and is particularly necessary to provide seamless mobility to delay-sensitive services. However, this feature also adds complexity to the Node B scheduler and the HARQ algorithm. Moreover, with soft handoff on E-DCH, the number of HSUPA physical channels to monitor for the UE is considerably higher.

In summary, HSUPA increases the uplink transmission throughput, decreases delays, and improves the coverage, compared to R99. HSUPA has a new uplink-dedicated transport channel called the enhanced dedicated channel (E-DCH). E-DCH is similar to DCH, with some improvements to support a shorter sub-frame and techniques such as hybrid HARQ and Node B scheduling. Because of the uplink configuration, HSUPA is more complex than HSDPA, and more signaling consuming. The HSUPA scheduler controls the radio resource less tightly than does the HSDPA scheduler.

USIM and UICC

It is possible to access a GSM network using a SIM card, as indicated earlier, and a UMTS network using a USIM card. A UICC (universal integrated circuit card) provides physical support within embedded or removable memory, in addition to an increasing number of other applications. The ETSI Smart Card Platform, SCP (www.etsi.org), and ISO (www.iso.org) develop and maintain standards and uniform platforms.

SPA+ and the Road to LTE

3GPP continued standardization work beyond HSUPA on multiple fronts, notably on the definition of an OFDM-based broadband wireless access technology LTE (long-term evolution) and on an all-IP network evolution SAE (system architecture evolution) as well as on a study of HSPA enhancements (HSPA+). The latter (Rel7) provides direction with respect to technology enhancements (e.g., advanced antenna technology) to operators interested in further improving their HSPA networks.

The road to LTE is discussed later in this chapter.

1.6.2 3GPP2 Radio Access Standards Evolution

Figure 1-17 shows a high-level view of the evolution of 3GPP2 radio-access technology. Some details of the cdmaOne to cdma2000 evolution are provided in this section, while UMB is introduced in the section, "Beyond 3G."

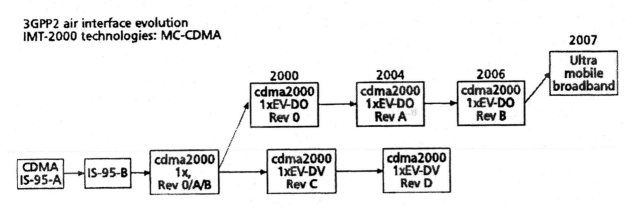

Figure 1-17: The Evolution of 3GPP2 Access-Technology Standards

1.6.2.1 cdmaOne and cdma2000-1X

cdmaOne was the first commercial CDMA network based on ANSI-95 standards. The cdma2000-1X doubled the spectral efficiency, as shown in Figure 1-18. The spectral efficiency here is defined by the sector capacity (in Mb/s) for both forward link (FL, base-to-mobile station) and reverse link (RL, mobile-to-base station) divided by the total required bandwidth (in MHz). The spectral efficiency was significantly enhanced by definition of the enhanced (data-optimized) version of cdma2000-1xEV-DO Rev. 0, and then Rev. A.

A cdmaOne network has pilot, sync, and paging control channels on the forward link and an access-control channel on the reverse link. Those control channels support forward- and reverse-link traffic channels. The forward-link channels are orthogonally covered by a Walsh function, and then spread by a quadrature pair of pseudo-random noise (PN) sequences at a fixed chip rate of 1.2288 Mc/s. After the spreading operation, the in-phase (I) and quadrature (Q) signals are applied to the inputs of the I and Q baseband filter. The (QPSK) modulated signals are then sent over the air, as shown in Figure 1-19.

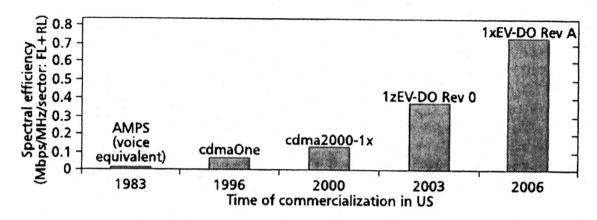

Figure 1-18: Evolution of Spectral Efficiency in 3GPP2

The pilot channel is an unmodulated all-zero signal transmitted continuously by each CDMA base station to provide mobile stations with timing for initial system acquisition, a phase reference for coherent demodulation, and a measure of signal strength for handoff decisions. The sync channel is a modulated spreading signal to transmit the sync channel message, which conveys key information to mobile stations, such as protocol revisions, the system identification number (SID), system time, pilot PN sequence offset index, and others. The paging channel sends control information to CDMA mobile stations that have not been assigned to a traffic channel. Forward traffic channels are used by the base station to pass user data and signaling messages to mobile stations while on a call. Forward traffic channels can operate at different rates depending on the service options supported.

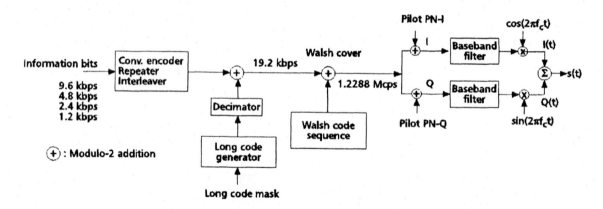

Figure 1-19: Block Diagram of Basic cdmaOne Forward-Link Transmission

When the mobile station is turned on, it first completes pilot acquisition and changes its de-spreading to the sync channel to receive sync channel messages. Mobile stations use sync information to synchronize their system timing to CDMA system time. Then, mobile stations perform system registration initialization and begin to monitor the paging channel. Once a mobile station is paged, or attempts to access the network, the mobile will switch to the traffic channel to start communication.

The reverse-link transmission is achieved with 42-bit PN codes, referred to as long codes. Each subscriber uses a unique long code. Walsh codes are used as modulation symbols, as illustrated in Figure 1-20.

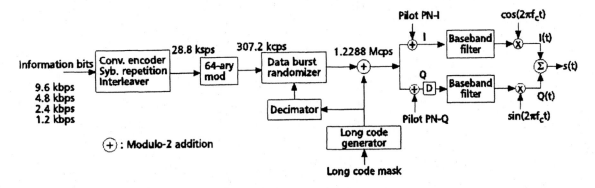

Figure 1-20: Block Diagram of Basic cdmaOne Reverse-Link Transmission

Access channels are used by mobile stations not engaged in a call to send messages to the base station. Typical messages sent include registration and call-origination messages, and responses to requests or to orders received from the base station. Reverse traffic channels are used by mobile stations while on a call to transmit user data and signaling information to the base station. The reverse-traffic channels can operate at different rates depending on the service options supported.

cdma2000-1X maintains the basic cdmaOne traffic channel structure and expands them to forward and reverse fundamental channels. It introduces forward and reverse supplemental channels, as well as associated forward and reverse dedicated control channels. Variable data rates in cdma2000 are achieved through variable spreading rates from 4 to 256 instead of the fixed 128 in cdmaOne. Both forward-link and reverse-link support up to 153.6 kb/s physical link data rate. A reverse-link pilot channel is added to support a coherent receiver on the reverse link. Powerful convolution and turbo codes that are specified double the voice capacity compared to cdmaOne, as well as support 144 kb/s packet data applications. The control channels are also upgraded to enhanced access-control channels and forward and reverse common-control channels with higher rates and shorter frames. A quick paging channel and a broadcast channel are introduced to improve call setup time and battery life.

Power control is essential for a CDMA system to control interference and ensure reliable system operation. The link-level radio power management is specified in the standards. It consists of forward-link and reverse-link power control and soft/softer handoff. The link-level power management determines the sector capacity. The two-fold increase in capacity from cdmaOne to cdma2000-1x was largely due to the introduction of fast forward-link power control. The forward-link power-control update rate increased from around 50 Hz in cdmaOne based on control messages, to a dedicated control channel at 800 Hz in cdma2000. Reverse-link power management has both open-loop and closed-loop power control. The open loop allows a mobile station to set its transmit power based on the strength of the power received on the forward link, while closed loop supports power control to a desired reverse-link traffic channel Eb/No, directed by the base station. These mechanisms ensure that the mobile and base stations will transmit at a minimal power level.

Soft and softer handoffs are also defined in CDMA standards to ensure make-before-break communication (as discussed earlier). Soft handoff refers to handoff between two adjacent sectors belonging to different cell sites. Softer handoff is handoff between two adjacent sectors of the same cell site. Although soft handoff improves voice quality and handoff performance, it ties up multiple channel receivers, and decreases capacity.

cdma2000-1x has also introduced various bearer service profiles to support voice-only, data-only, mixed-voice and data services over traffic channels. Medium access layer, link access layer, and upper layer protocol combine to provide an efficient way to transmit bursty packet-data traffic with differentiated QoS capability. New network elements were also introduced to address data mobility, IP addresses, packet data routing, authentication, authorization, and accounting, as illustrated in Figure 1-21.

1.6.2.2 cdma2000-1xEV-DO (Rev. 0 and Rev. A)

Cdma2000-1xEV-DO is designed for high-data-rate, flexible-latency, and high- quality applications. The 1xEV-DO high-speed packet-data network is an integral part of the cdma2000 family of standards. The 1xEV-DO has the same spectrum bandwidth as that of cdma2000-1x and is spread by the same PN sequences as cdma2000-1x. Hybrid mode supports smooth seamless handoff between 1xEV-DO and cdma2000-1x. An integrated 1xEV-DO and cdma2000-1x network architecture is shown in Figure 1-21.

The architecture provides a spectrally efficient means for both delay-sensitive (e.g., conversational) and delay-tolerant data services. The integrated network can share packet-data network elements, such as packet data serving node (PDSN), foreign agent (FA), authentication, authorization and accounting (AAA) and home agent (HA). Only a radio network controller (RNC) and dedicated 1xEV-DO base station modem cards are needed to upgrade a deployed cdma2000-1x network.

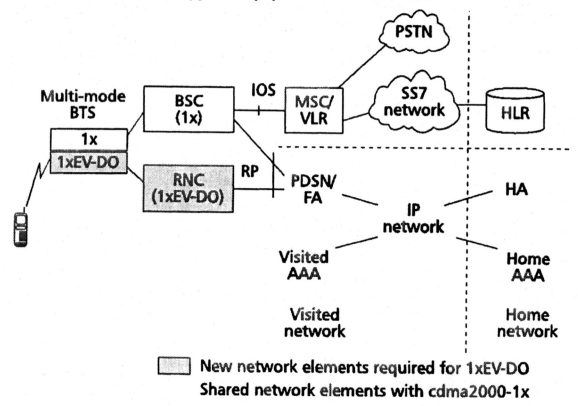

Figure 1-21: Architecture of a cdma2000 and 1xEV-DO Network

The cdma2000-1xEV-DO Rev. 0 network is dedicated to packet-data traffic, and optimized for maximum data throughput. The system takes advantage of the burst and delay-tolerant nature of data to increase throughput efficiency. High spectral efficiency in 1xEV-DO is achieved on a shared full-power and full-code forward-link fat pipe through adaptive modulation and coding (AMC), fast scheduler, and hybrid-ARQ (H-ARQ).

The forward traffic channel is a common shared channel for all access terminals and their control signals. Accordingly, the modulation and coding are adaptive to the channel carrier-to-interference ratio (CIR). QPSK, 8-PSK, and 16-QAM modulations with various turbo coding are used to achieve data rates ranging from 38.4 kb/s to 2.457 Mb/s in a 1.25-MHz channel, as shown in Table 1-2. The transmission data rate to each terminal is adapted to the RF channel condition and decided by the terminal. Therefore, user under good RF conditions will transmit at a higher data rate than users in poor RF conditions. The data rate control (DRC) channel on the reverse link provides the fast feedback of the data rate and best serving sector at a 1.667-ms slot interval. Fast cell selection supports virtual soft handoff. Only the best serving sector is transmitting to the terminal at a time. This reduces interference and consumption of resources.

Data Rate (kb/s)	Number of Slots	Code Rate	Modulation	CIR (dB)
38.4	16	1/5	QPSK	-11.5
76.8	8	1/5	QPSK	-9.7
153.6	4	1/5	QPSK	-6.8
307.2	2	1/5	QPSK	-3.9
614.4	1	1/3	QPSK	-3.8
307.2	4	1/3	QPSK	-0.6
614.4	2	1/3	QPSK	-0.8
1228.8	1	1/3	QPSK	1.8
921.6	2	1/3	8-PSK	3.7
1843.2	1	1/3	8-PSK	3.8
1228.8	2	1/3	16-QAM	7.5
2457.6	1	1/3	16-QAM	9.7

Table 1-2: Modulation and Coding Schemes of -1xEV-DO Rev. 0

Fast scheduler is enabled by channel state information feedback from the DRC channel as well. The scheduler takes advantage of rapidly changing fading channel characteristics, and serves users when they have better than average channel CIR. This achieves diversity gain. The transmission time and duration is decided by the base station based on both the history of the transmission and the usage pattern. This adaptive data scheduler can support fairness and differentiation among users.

Cdma2000 1xEV-DO Rev. A is designed for versatile multimedia services, both delay-sensitive and delay-tolerant applications. Major improvements over EV-DO Rev. 0 are increased data rates, improved reverse link, and a multiple-flow packet application to support delay-sensitive applications with more symmetric data-rate requirements, such as voice and video telephony.

EV-DO Rev. A increases the forward link data rate to 3.072 Mb/s by increasing the physical layer packet size from 4096 bits at 2.457 Mb/s in Table 1-2 to 5120 bits. The reverse-link data rates are expanded from 4.8 kb/s to 1.843 Mb/s. Higher data rates and better packing efficiency improve the forward- and reverse-link spectral efficiency, and enable VoIP capability, among others.

The EV-DO Rev. A reverse-link performance is improved through hybrid-automatic repeat request (H-ARQ) and two transmission modes, while EV-DO Rev. 0 has a reverse-link structure similar to that of cdma2000-1x. H-ARQ achieves higher effective data throughput by earlier termination of multiple sub-packet transmissions. H-ARQ exploits power-control imperfections due to radio-channel and interference variations using a-posteriori channel feedback. Two transmission modes are high capacity (HiCap) and low latency (LoLat). HiCap has a termination target of 4 sub-packets, while LoLat's termination target is 2 to 3 sub-packets. HiCap achieves higher sector-capacity gain with lower required bit-energy-to-noise-density ratio. LoLat minimizes transmission delay and achieves a lower packet-error rate.

A multiple-flow packet application supports spectral efficiency and latency trade-off for each application flow. A service provider can deliver various multimedia applications with distinctive flow characteristics for each radio link protocol (RLP) flow over EV-DO Rev. A. 1xEV-DO Rev. A base stations manage radio resources to meet user throughput, latency and packet-loss requirements and to ensure end-to-end QoS. Forward-link scheduler, reverse-link traffic power and medium-access control, call-admission control, backhaul QoS support and overload control, together form an integrated and effective resource-management system.

Forward-link QoS is managed by a RF scheduler that has a queue for each RLP flow. It assigns the packets in accordance with the needs of each flow QoS profile to perform expedited forwarding (EF), assured forwarding (AF) and best effort BE) resource allocation. The voice flow is given the highest delay and jitter-sensitive EF priority to ensure voice quality. Background file download is assigned BE flow.

Reverselink QoS is controlled by a token bucket scheme to ensure that average and peak resources utilized by each reverse traffic channel are less than or equal to the limitations imposed by RAN. The limitations are specified by traffic-to-pilot ratios (T2P). LoLat has higher T2P ratios for the first 2 to 3 sub-packets. HiCap has the same T2P ratios for all 4 sub-packets, as illustrated in Figure 1-22.

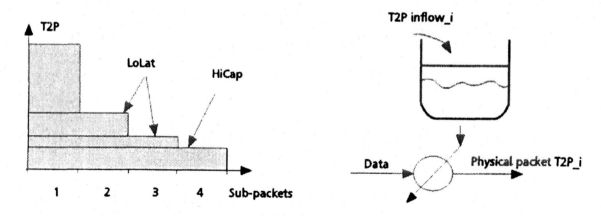

Figure 1-22: Reverse-Link Resource Management for QoS

1xEV-DO Rev. A has improved physical, MAC, and higher layer network efficiency and latency performance for multimedia services. The spectral efficiency for packet-data applications has reached 0.72 Mb/s/MHz/sector. 1xEV-DO can also support voice user capacity of 35 Erlangs at a 1.25-MHz bandwidth, and it can also support a wide range of multimedia services.

1.6.3 Broadcast/Multicast Distribution over Cellular

Broadcast/multicast services such as mobile TV are being increasingly introduced as multimedia content applications are enabled by evolving wireless systems. The distribution systems may be:

- Satellite ("S") or terrestrial ("T")
- Cellular network or overlay access

These systems leverage sophisticated technologies for such functions as coding, modulation, error correction, media formatting and presentation, content protection and management, time-slicing, radio-networking, and data-casting. Typically, frequencies below 800 MHz have high capabilities for non-line-of-sight reach and indoor penetration for an application such as over-the-air TV; though higher frequencies are also used, e.g., for high-capacity direct access. An end-to-end block diagram indicating the wireless-access component is shown in Figure 1-23.

The Open Mobile Alliance has worked to develop standards for service enablers, including unified broadcast/multicast service enabler standards, with adaptation to a number of distribution systems.

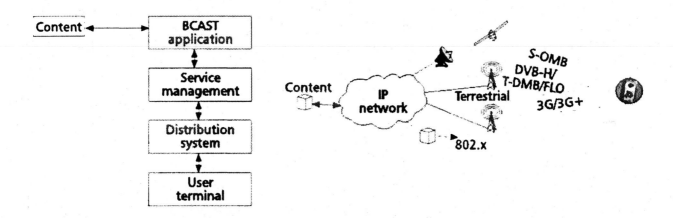

Figure 1-23: A Simplified End-To-End Broadcast/Multicast system

Examples of mobile broadcast/multicast distribution systems, integrated within or overlaid on top of mobile cellular systems include, but are not limited to:

- MBMS (3GPP's multimedia broadcast/multicast service)–www.3gpp.org
- BCMCS (3GPP2's broadcast and multicast services)–www.3gpp2.org
- DVB-H (digital video broadcast–handheld)–www.dvb-h.org
- DMB (digital multimedia broadcasting)–http://eng.t-dmb.org
- MediaFLO (forward-link only technology-from Qualcomm)–www.mediaflo.com

1.7   Local, Personal and Near-Field Communications

Although some access technologies compete with each other as an optimal choice for an application, their complementary nature to serve the user requirements within a dynamic and cooperative communication environment must be appreciated. Serving the user across different domains and for different applications requires a seamless use of one of more access technology within a personal, local, or wide area.

1.7.1 Wireless Local Area Network (WLAN)

The WLAN protocol is defined by the IEEE 802.11 standard, which was originally published in 1997 and updated with a main revision in 2007. A letter next to 802.11 notes the various versions of WLAN. All are amendments to the original IEEE 802.11 standard. The following standards and amendments exist within the IEEE 802.11 family:

- IEEE 802.11—The original WLAN standard, 1 Mb/s and 2 Mb/s, 2.4 GHz, RF and IR standard (1997); all the others listed below (defined, in progress or proposed) are amendments to this standard, except for Recommended Practices 802.11F and 802.11T.
- IEEE 802.11a—54-Mb/s, 5-GHz standard (1999)
- IEEE 802.11b—Enhancements to 802.11 to support 5.5 Mb/s and 11 Mb/s (1999)
- IEEE 802.11c—Bridge operation procedures (2001)
- IEEE 802.11d—International (country-to-country) roaming extensions (2001)
- IEEE 802.11e—Enhancements: QoS, including packet bursting (2005)
- IEEE 802.11g—54-Mb/s, 2.4-GHz standard (backwards compatible with .11b) (2003)
- IEEE 802.11h—Spectrum-managed 802.11a (5 GHz) for European compatibility (2004)
- IEEE 802.11i—Enhanced security (2004)
- IEEE 802.11j—Extensions for Japan (2004)
- IEEE 802.11-2007—New release of the standard with amendments a, b, d, e, g, h, i & j (July 2007)
- IEEE 802.11k—Radio resource measurement enhancements
- IEEE 802.11l—Reserved
- IEEE 802.11m—Maintenance of the standard. Recent edits became 802.11-2007(ongoing)
- IEEE 802.11n—Higher throughput improvements using MIMO (multiple-input multiple-output antennas)
- IEEE 802.11o—Reserved
- IEEE 802.11p—WAVE(Wireless access for the vehicular environment)
- IEEE 802.11q—Reserved
- IEEE 802.11r—Fast roaming
- IEEE 802.11s—ESS (Extended service set) for mesh networking
- IEEE 802.11T—Wireless performance prediction (WPP): test methods and metrics (recommendation)
- IEEE 802.11u—Interworking with non-802 networks (for example, cellular)
- IEEE 802.11v—Wireless network management
- IEEE 802.11w—Protected management frames

- IEEE 802.11x—Reserved
- IEEE 802.11y—3650-3700 MHz operation in the U.S.
- IEEE 802.11z—Extensions to direct link setup (DLS) (Aug. 2007-Dec. 2011)

Today's commercial WLAN products are mostly based on 802.11a, 802.11b, 802.11g, and the recently introduced 802.11n protocols. While 802.11a was the first wireless networking standard, the first standard broadly implemented was 802.11b, followed by 802.11g and now 802.11n. The standards c–f, h, and j are service amendments and extensions or corrections to previous specifications. The new 802.11i was introduced to fix the security weaknesses of the WEP (wireless encryption protocol) that came with the original 802.11 standard.

Protocol Description

As with other 802.x protocols, 802.11 deals with the data link and physical layers of the OSI (Open System Interconnection) protocol stack. The data link layer functions are covered by the MAC (media access control) protocol of the 802.11 standard. Figure 1-24 shows the relevance of 802.11 in the OSI protocol stack.

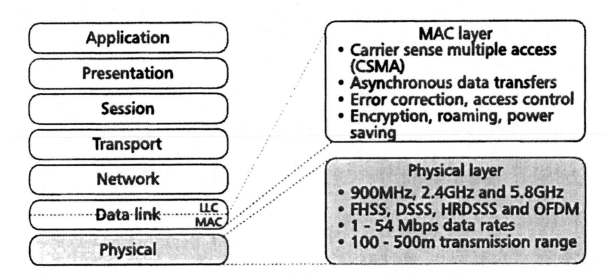

Figure 1-24: The IEEE 802.11 Protocol Stack

MAC Layer

Some of the major features of the MAC layer are:

- CSMA/CA (carrier sense multiple access with collision avoidance) is its basic access mechanism. It does not transmit if the medium is thought to be occupied by other users, and only transmits when the medium is believed to be available for communication. The transmitting station is allowed to transmit upon sensing the free medium for a specified time called the DIFS (distributed inter-frame space). The receiving station will then check the CRC (cyclic redundancy check) of the received packet and send an ACK (acknowledgement). The receipt of an ACK by the sender station ensures that no collision has taken place. If the sender does not receive an ACK within a given time, the packet is retransmitted. The sender will discard the packet if the number of retransmissions reaches a maximum limit.

- The 802.11 standard provides a mechanism for the "virtual carrier sense" with an acknowledgement signaling protocol. A station willing to transmit a packet first sends a short packet called an RTS (request to send) which will contain the source, destination and duration of the following transaction. If the receiving station is free, it will send a response with another short packet called the CTS (clear to send) that denotes the sender can now transmit its message. All stations receiving either the RTS or the CTS will set their NAV (network allocation vector) for the given duration and use this information (along with the physical carrier sense) to sense the medium.
- A simple send-and-wait algorithm addresses the fragmentation and reassembly functions. The transmitting station is not allowed to send any new fragment until it receives an ACK for the fragment sent, or it decides that the fragment was retransmitted too many times.
- 802.11 uses an exponential back-off algorithm to resolve contention between multiple stations attempting to access the medium simultaneously. This method requires each station to wait for a random number of slots before attempting to access the medium again.

Physical Layer

Here are some of the highlights of the 802.11 physical layer.

- An 802.11 protocol is allowed to operate in the unlicensed 915-MHz, 2.4-GHz, and 5-GHz ISM (Industry, Science, and Medical) bands.
- 802.11a protocol operates on the 5-GHz U-NII (unlicensed national information infrastructure) band with 12 non-overlapping channels. It uses a 52 subcarrier OFDM (orthogonal frequency division multiplexing)-based modulation.
- Both 802.11b and 802.11g protocols use the 2.4-GHz ISM spectrum with three non-overlapping channels. Since ISM spectrum is free, 802.11b and g equipment may occasionally suffer interference from other household devices such as microwave ovens and cordless telephones.
- 802.11b protocol uses DSSS (direct sequence spread spectrum) modulation while 802.11g uses OFDM-based modulation.
- 802.11n protocol builds on previous 802.11 standards by adding MIMO (multiple-input multiple-output) to the physical layer. MIMO uses multiple transmitter and receiver antennas to improve system performance. 802.11n operates in both 2.4-GHz and 5-GHz spectrum.

Table 1-3 provides a brief comparison of physical layer characteristics. More information is available at

http://en.wikipedia.org/wiki/IEEE_802.11

Protocol	Release date	Operating frequency (GHz)	Realistic throughput (Mb/s)	Max. data rate (Mb/s)	Modulation technique	Indoor range radius (meters)	Outdoor range radius (meters)
Legacy	1997	2.4	0.9	2		~20	~100
802.11a	1999	5	23	54	OFDM	~35	~120
802.11b	1999	2.4	4.3	11	DSSS	~38	~140
802.11g	2003	2.4	19	54	OFDM	~38	~140
802.11n	June 2009 (est.)	2.4 and 5	74	248	DSSS or OFDM	~70	~250

Table 1-3: Physical Layer Characteristics of 802.11 Protocols

Network Architecture

Figure 1-25 shows a typical reference architecture of an 802.11-based WLAN. The network is divided into regions called basic service sets (BSS), each of which is controlled by a radio base station called an AP (access point). The boundary of each BSS is defined by the range of the AP for providing wireless connectivity to the MSs (mobile stations) within that BSS. An Ethernet-based backbone network typically handles the interconnection between multiple BSSs. However, the BSSs can be connected via a wireless backbone network as well. The interconnected BSS area is called an ESS (extended service set) in the standard.

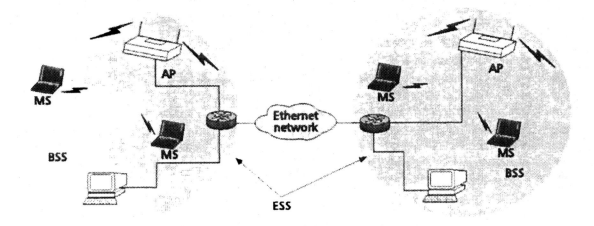

Figure 1-25: An IEEE 802.11 WLAN Network Reference Model

An independent BSS (IBSS) is a network architecture where the MSs talk directly to each other and are not connected to a wired network (e.g., the Internet and/or a private IP-based network). These MSs can only talk to each other and build their own LAN (local area network).

In an IBSS, an MS can only communicate with another MS if they are in direct radio contact. There is no central node (e.g., an AP) where all the MSs can connect, and the connections between the MSs are established as needed. In such a case, the IBSS is also called an ad-hoc network, which is usually small and short-lived (relative to a BSS or an ESS). For example, in a conference setting, the presenter and the participants could exchange their files without having to log onto the conference center's private network.

1.7.2 Wireless Personal Area Network (WPAN)

The wireless personal-area network (WPAN) enables wireless connectivity and communications between devices and sensors within a few meters to tens of meters of each other. They typically operate at low power and have capabilities ranging from low to high data rates. Examples include IrDA, Bluetooth, Zigbee, and UWB (ultra-wideband).

The sub-working groups within IEEE 802.15 cover a range of capabilities, from simple and low-rate (e.g., Zigbee) interaction of devices and sensors (802.15.4), to high data-rate definition (802.15.3) for such applications as short-range multimedia imaging or vehicular communications. The 802.15.1 has been a cooperative effort with the Bluetooth special interest group (SIG) formed in 1998, www.bluetooth.org). Many devices, particularly mobile user terminals, are equipped with Bluetooth.

The IEEE 802.15 WPAN task groups are listed in Table 1-4 (http://standards.ieee.org).

IEEE 802.15 WPAN Task Group	Scope
802.15.1	WPAN/Bluetooth
802.15.2	Co-existence
802.15.3	High-rate WPAN
802.15.4	Low-rate WPAN
802.15.5	Mesh networking
802.15.6	Body area networks (BAN)

Table 1-4: IEEE 802.15 WPAN Task Groups

Ultra Wideband (UWB) is a technology that uses a large bandwidth, potentially shared with other users, to transmit information at low power. It is intended for a range of applications, notably high-data-rate personal-area networking. While technical, spectrum, and regulatory concerns have been debated and the question of co-existence is expected to be further addressed, UWB has received a flurry of research as a technology with a great deal of prospects.

Important parameters of a wireless PAN naturally include such attributes as transmit power, range, data rate, spectrum, co-existence/interference, configuration/set-up, pairing, and mesh networking.

1.7.3 Near-Field Communication & RFID

Radio-frequency identification (RFID) is a mechanism for reading data stored on tags from distances of centimeters to tens of meters away. Interest in RFID is growing, particularly for enterprise applications such as inventory tracking.

The RFID tags may be passive, semi-passive, or active. Passive tags do not have a power source and are activated by an RFID reader.

Near-field communication (NFC) allows short-range wireless connectivity, typically within centimeters, enabling such applications as secure and reliable mobile payment and ticketing (in a crowd of users and in the presence of other applications). It is worth noting that mobile (2D) barcodes are also finding a great market in mobile commerce, among other applications, typically using optical (camera) scanning of information (e.g., on a URL).

NFC is compatible with RFID and extends its capabilities. It can co-exist with WPAN, e.g., to configure or pair with it. Despite some overlap between the two, the distinction between NFC and WPAN are self-evident in terms of such attributes as range, power, data rate, set-up time, topology, and applications. As with other access mechanisms, however, their complementary nature must be appreciated to be leveraged effectively.

1.7.4 Femto Cells for Home Networks

A femto cell is an independent extension of a wide-area network. It helps improve the coverage within a local (home) area to serve a limited number of users. The goal is for the cell to have many features of the wide-area network but with minimal impact (e.g., in terms of performance, capacity, and efficiency, etc.).

The features and attributes of a macrocellular network that may be supported by femto cells (depending on the service terms and policies) include:

- Services at quality comparable to main network
- Radio interfaces, signaling
- Mobile terminals (governed by the subscription's authorization policy and admission control)
- Mobility management, handoff, operational management, security

The overall architecture is intuitively simple. Each femto cell has a base station at low power connected in a local (home) coverage serving authorized mobile devices, which may move in and out of the cell. The femto cell home network is connected through an existing (IP) broadband connection to the Internet and a cellular network as shown in Figure 1-26.

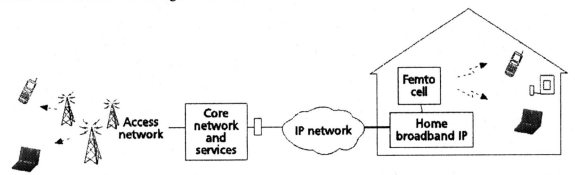

Figure 1-26: A Femto Cell As an (Independent) Extension of a Wide-area Network

The macrocellular core network must be able to support a large number of femto cells without a noticeable compromise in efficiency, performance, or security.

1.7.5 Hybrid Systems

Wireless access technologies are increasingly designed to be compatible or inter-work. In addition, cooperative and hybrid (IP) sub-networks exist at the network level and provide the user terminal the ability to work across the sub-networks while maintaining service continuity. Examples include macrocellular networks and femto cells, 2G and 3G (and beyond) access environments, 3G and 802.x, etc.

For example, a hybrid network consisting of 3G wireless and WLAN technologies is more than a reality today, especially for data services. An easy way to connect these two disparate networks is by using MIP (mobile IP) protocol as the underlying glue. Since both 3G and WLAN access technologies use packet-based IP routing, the MIP protocol lends itself well to a seamless interaction between these networks.

MIP protocol uses an IP-in-IP tunneling method to forward the data packets to the FA (foreign agent) that serves as the point of attachment for a MS. Upon connecting itself to an access network—whether 3G or WLAN—the MS must register with a FA associated with that specific access network. The FA, in turn, then informs the HA (located at the home network) of the MS's current location (at the serving network). This enables the HA to create a location map against the MS's IP address, which is topologically the same as the HA's routing domain (i.e., the home network). Any data packet destined for the MS is then first intercepted by the HA at the home network and subsequently forwarded by the HA to the current FA at the serving network. The FA retrieves the original packet by discarding the outer header of the IP-in-IP tunnel (created earlier by the HA) and forwards it to the MS.

Figure 1-27 shows a hybrid network consisting of 3G and WLAN networks. The 3G and WLAN access components, classified as serving networks have FAs associated with them. The home network, on the other hand, houses the HA component. As the MS moves between 3G and WLAN access networks, it registers with the new FA so that the HA can forward packets accordingly.

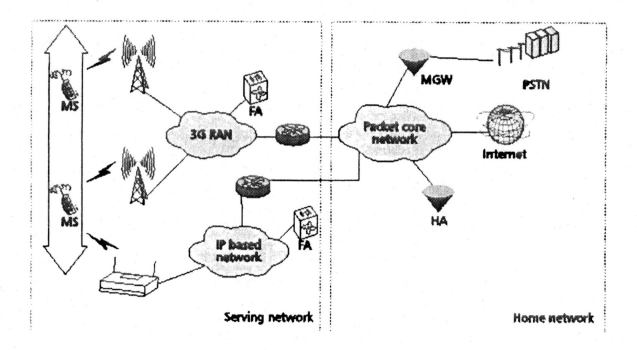

Figure 1-27: A 3G and WLAN Hybrid Network Using Mobile IP

1.8   Beyond 3G and Future Trends

1.8.1 Motivation and Design Objectives

Wireless evolution beyond the third generation has attracted a great deal of interest in standardization and regulatory entities. Efforts are being made to shape design goals and specify technical requirements to serve an increasingly mobile world with ubiquitous connectivity, communication, and content on-the-go.

In the same framework, a number of global and regional initiatives have tried to create think tanks to help shape technical and strategic directions of systems beyond 3G. They take into account the latest developments in enabling wireless technologies and networking research and the current spectrum and regulatory landscape.

A number of technical challenges must be addressed, mainly associated with high data rates in a variety of propagation and traffic conditions, efficient resource allocation and interference management, realistic performance evaluation, overhead signaling, and system complexity.

With HSDPA defined in Rel 5 and HSUPA/Enhanced DCH introduced in Rel 6, 3GPP considerably enhanced the spectral efficiency of UMTS. Similarly, significant advances have been made in cdma2000 access technology standards, as discussed earlier. Further improvements have been introduced in subsequent releases and versions by 3GPP and 3GPP2. Mostly these have improved the bit rate and capabilities of 3G technologies (e.g., broadcast/multicast, QoS, VoIP, etc.). In parallel, concurrent technologies emerged, notably through the IEEE 802.16 standards (with significant contributions from the IT/computing community) coupled with much related work in the WiMAX Forum. This helped remarkably to inspire a new direction for the logical evolution of the mobile ecosystem.

In essence, the design goals for systems beyond 3G include:

- High data rate (speed) and capacity (traffic volume)
- Low latency (real-time and near-real-time), high performance, constant connectivity
- Spectral efficiency
- Cost/resource efficiency
- Flexible resource allocation, adaptive interference management, seamless user mobility, scalable spectrum
- Enhanced broadcast/multicast and end-to-end QoS
- Co-existence and inter-working with other access technologies

It is intuitively appealing that some of these goals, such as seamless mobility and inter-working, end-to-end QoS, etc., are tied to the evolution of the core packet network, which is moving in parallel toward a heterogeneous all-IP flat (or flatter) architecture. (Core network evolution is the subject of Chapter 2.)

Wireless access technologies beyond 3G leverage such sophisticated, efficient, and high-capacity technologies as OFDMA, MIMO, and advanced coding/high-level modulation (e.g., 16- or 64-QAM), among others.

Three remarkable candidates, LTE, UMB, and mobile WiMAX, are being considered for wireless access technologies beyond 3G.

1.8.2 Wireless Access Standards Beyond 3G

1.8.2.1 LTE

3GPP's work on the evolution of 3G started in 2004. A set of high-level requirements was identified, including reduced cost per bit, increased service provisioning, flexibility in use of existing and new frequency bands, simplified architecture, open interfaces, and reasonable terminal power consumption.

A feasibility study of the UMTS *long-term evolution* (LTE) started immediately. The objective was to develop a framework for the evolution of the 3GPP radio-access technology towards high-data-rate, low-latency, and packet-optimized radio access. The study focused on the following:

- A radio-interface physical layer (downlink and uplink) to support a flexible transmission bandwidth up to 20 MHz by introducing new transmission schemes and advanced multi-antenna technologies
- Radio interface layers 2 and 3 and signaling optimization
- UTRAN architecture, identifying the optimum network architecture and functional split among RAN network nodes
- RF-related concerns

A set of basic requirements resulted from the first part of the study (3GPP, March 2006):

- Peak data rate: Instantaneous downlink peak data rate of 100 Mb/s within a 20-MHz downlink spectrum allocation (5 b/s/Hz) and instantaneous uplink peak data rate of 50 Mb/s (2.5 b/s/Hz) within a 20-MHz uplink spectrum allocation
- Control-plane latency: Transition time of less than 100 ms from a camped state to an active state and transition time of less than 50 ms between a dormant state and an active state
- Control-plane capacity: At least 200 users per cell should be supported in the active state for spectrum allocations up to 5 MHz
- User-plane latency: Less than 5 ms in non-loaded condition (i.e., a single user with a single data stream) for small IP packets
- User throughput: Average downlink user throughput per MHz, 3 to 4 times that of HSDPA and average uplink user throughput per MHz, 2 to 3 times that of the enhanced uplink (Release 6)
- Spectral efficiency: Same improvement factors as for throughput
- Mobility: Performance should be optimized for low mobility, but high mobility across the cellular network should be maintained at speeds greater than 120 km/h (and up to 500 km/h depending on the frequency band)
- Coverage: The throughput, spectrum efficiency and mobility targets above should be met for 5-km cells, and with slight degradation for 30-km cells
- Enhanced multimedia broadcast multicast service (MBMS)
- Paired and unpaired spectrum arrangements
- Spectrum flexibility and scalability
- Co-existence and inter-working with 3GPP radio access technologies
- Architecture and migration: Single E-UTRAN architecture and optimized backhaul communication protocols
- Radio resource management requirements for enhanced end-to-end QoS support, efficient support for transmission of higher layers and load sharing and policy management across different radio access technologies
- Complexity optimization (with a limited number of options and no mandatory redundant features).

A number of multiple-access schemes have been considered, each with the objective of supporting both FDD and TDD operation. OFDMA was selected for the downlink, single carrier frequency division multiple access (SC-FDMA) for the uplink. OFDMA is attractive for the downlink not only for its increased spectral efficiency but also for its low complexity. This is even more crucial when MIMO is used and the inherent orthogonality in transmission that eliminates intracell interference is considered. While OFDMA could have been an attractive multiple-access scheme for the uplink, OFDM signals have a high peak-to-average power ratio (PAPR). Large PAPR requires a large back-off in the power amplifier and decreases coverage area, resulting in poor cell-edge performance. To address the PAPR issue while preserving the advantage of the "orthogonality" of OFDM, and therefore to combine the benefits of OFDMA and single-carrier transmission, single-carrier FDMA (SC-FDMA) was chosen.

Adaptive modulation and coding are used in the downlink to adapt the data rate and QoS requirements of the users to the channel quality. Various coding schemes have been considered, such as duo-binary turbo codes, rate compatible/quasi cyclic LDPC codes (RC/QCLDPC), concatenated zigzag LDPC codes, turbo single-parity-check (SPC) LDPC codes and shortened turbo codes with the insertion of temporary bits. They have different performance merits for different block sizes and complexity assumptions.

To address the interference-limited scenarios caused by universal frequency reuse, interference mitigation can be deployed in the form of:

- Inter-cell interference randomization by deploying cell-specific scrambling, cell-specific interleaving, or random frequency hopping.
- Interference avoidance, by means of fractional frequency reuse: Users are allocated in multiple classes according to their distance from the cell center and the bandwidth allocation patterns assigned to different user classes.
- Interference cancellation by multiple antenna processing.

Multiple-antenna techniques are considered in 3GPP LTE to increase data rates and improve link quality by using spatial multiplexing and diversity gains. As these schemes are deployed in combination with the packet scheduler, performance optimization needs to be studied. This must take into account both spatial and multi-user diversity gains at a system (multi-cell) level to include realistic interference modeling. In the single-user MIMO case, for example, in a per-antenna rate-control (PARC) scheme, spatial multiplexing is implemented to improve the link rate by transmitting parallel data streams to a single user. In a multi-user MIMO case, multiple spatial streams are transmitted to different users to increase system throughput. Space division multiple access (SDMA) can be realized through sectorization, multi-user beam forming or multi-user MIMO precoding. Decisive factors in the realistic assessment of candidate MIMO approaches are the system-level performance under realistic channel and interference modeling assumptions, and the required channel-state information, overhead signaling, and pilot design constraints.

In summary, LTE is a broadband wireless access technology (with flexible capabilities) defined by 3GPP in Release 8, to enable high data rates, high capacity, and reduced latency, among other features, and with significant enhancements in spectral and resource efficiencies. It supports both TDD and FDD schemes and enhanced broadcast/multicast, with a transmission bandwidth ranging from 1.25 MHz to 20 MHz. An OFDMA mechanism is used in downlink and single-carrier FDMA in uplink. In addition, such elements as sophisticated antenna systems (e.g., 2x2 or 2x4 MIMO) and coding/modulation schemes (e.g., QPSK, 16-QAM, 64-QAM) will lead to an evolution of LTE products. LTE is expected to support service continuity and handoff across a variety of 3GPP and non-3GPP access technologies.

1.8.2.2 UMB

The 3rd Generation Partnership Project 2 (3GPP2) standards body is engaged in an activity similar to the 3GPP LTE effort, aiming at moving the cdma2000 high-rate packet-data (HRPD) 3G system into the next generation. The 1x EV-DO Rev. C standard, now known as ultra mobile broadband (UMB), was specified and released in 2007. It is based on OFDMA, with sophisticated control and signaling mechanisms, radio resource management (RRM), adaptive interference management, and advanced antenna techniques, such as MIMO, SDMA, and beam-forming.

UMB also addresses the need for advanced mobile broadband services with efficient delivery. It supports inter-technology handoffs and seamless operation with cdma2000 systems.

The basic target requirements for UMB are:

- High-speed data: Peak rates of 288 Mb/s (with 2x20-MHz FDD, 4x4 MIMO) download and 75 Mb/s (with 2x20-MHz FDD, 1x2 MIMO) upload.
- Increased data capacity: Ability to deliver both high-capacity voice and broadband data in all environments; fixed, pedestrian, and fully mobile in excess of 300 km/h.
- Low latency: An average latency of 14.3 ms over-the-air to support VoIP, push-to-talk and other delay-sensitive applications with minimal jitter.

- Increased VoIP capacity: Up to 1000 simultaneous voice over IP (VoIP) users within a single sector, a 20-MHz bandwidth in a mobile environment without degrading concurrent data-throughput capacity.
- Coverage: Large wide-area network coverage areas equivalent to existing cellular networks, with either ubiquitous coverage for seamless roaming or non-contiguous coverage for hot-zone applications.
- Mobility: Robust mobility support with seamless handoffs.
- Converged access network, which is an advanced IP-based radio access network (RAN) architecture being developed by 3GPP2 to support multiple-access technologies and advanced network capabilities, such as enhanced QoS, with fewer network nodes and lower latencies.
- Multicasting: Support for high-speed multicast of rich multimedia content.
- Transmission bandwidth: Deployable in flexible bandwidth allocations between 1.25 MHz and 20 MHz.

The UMB air interface supports both TDD and FDD modes and several coding schemes, including turbo and (as an option) LDPC. OFDMA is used for uplink and downlink traffic, while both OFDMA and CDMA are used for uplink control channel transmission. Orthogonal access schemes (such as OFDMA) eliminate intracell interference but require centralized assignment of resources to users, thus requiring additional signaling and possibly adding to the overall latency. On the other hand, non-orthogonal access schemes can take advantage of "soft capacity" and do not require explicit signaling for resource allocation. However, they involve receivers that are more complex and control methods to achieve a link performance similar to that of orthogonal schemes. For this reason, for the UMB uplink control channel (which is by nature bursty, delay-sensitive, and with small payloads), a hybrid scheme is used composed of both OFDMA and a non-orthogonal multi-carrier CDMA (MC-CDMA) component. Both uplink and downlink support a variety of modulation mechanisms.

UMB downlink supports two MIMO schemes:

- Single codeword (SCW) MIMO with closed loop rate and rank adaptation: A single packet is encoded and sent on a number of spatial beams for rate adaptation. The initial UMB design will be based on SCW MIMO.
- Multi-codeword (MCW) MIMO with per-layer rate adaptation: Multiple packets are encoded separately before spatial multiplexing.

UMB also supports adaptive interference management for increased cell-edge data rates. Furthermore, flexible resource allocation is optimized for low overhead signaling through the classification of assignments as "persistent" or "non-persistent." Persistent assignments eliminate bandwidth-request latency for delay-sensitive applications, whereas non-persistent assignments are best suited for best-effort traffic. Seamless handoffs and advanced QoS mechanisms enable enhanced performance and support for real-time services.

1.8.2.3 Mobile WiMAX

Mobile WiMAX is a broadband wireless technology that supports both delay-sensitive (conversational or interactive) and delay-tolerant (non-real-time) applications. Mobile WiMAX has made significant advances as a candidate for the wireless space beyond 3G.

The air interface of the mobile WiMAX is based on an OFDMA (orthogonal frequency division multiple access) protocol as specified in IEEE 802.16-2004. The mobility enhancements of the IEEE 802.16e mobile amendment provide improved multi-path performance in non-line-of-sight environments and support for scalable-channel-bandwidth operation from 1.25 to 20 MHz (as outlined by the WiMAX Forum in February 2006).

The Mobile Technical Group (MTG) in the WiMAX Forum has defined several mobile WiMAX profiles to come up with various capacity-optimized and coverage-optimized configurations. Initial mobile WiMAX profiles, however, will cover 5-, 7-, 8.75-, and 10-MHz channel bandwidths for licensed worldwide spectrum allocations in the 2.3-, 2.5- and 3.5-GHz frequency bands.

The Network Working Group (NWG) of the WiMAX forum defines the higher-level networking specifications for mobile WiMAX systems beyond the IEEE 802.16 standard. The combined effort of IEEE 802.16 (http://standards.ieee.org) and the WiMAX Forum (http://www.wimaxforum.org) has helped define the end-to-end system solution. Mobile WiMAX:

- Supports peak data-rate targets of up to 63 Mb/s on the DL and of up to 28 Mb/s in a 10-MHz channel on the UL. This requires use of MIMO antennas, flexible sub-channelization schemes, efficient MAC frames, and enhanced MCS (modulation and coding schemes).
- Dynamically allocates RAN resources based on service flows, which is another advanced feature. Service flows enable the RAN to utilize its resources efficiently based on such QoS requirements as low latency.
- Supports disparate allocations of spectrum based on requirements, and scales within a variety of channelization schemes between 1.25 MHz and 20 MHz.
- Provides security by implementing EAP (extensible authentication protocol)-based authentication, AES-CCM (advanced encryption standard–CTR mode with CBC-MAC)-based authenticated encryption, and other message-authentication codes.
- Supports SIM/USIM (subscriber identity module/universal subscriber identity module) cards, smart cards, digital certificates and username/password associated with EAP methods for credential management.
- Features an optimized handoff scheme with a 50-ms mute time, because fast handoff is critical for supporting delay-sensitive applications.

Network Architecture

Figure 1-28 shows the network reference architecture for mobile WiMAX. It depicts the key normative reference points R1 to R5. Each of the entities, MS, ASN and CSN, represent a group of functional entities, each of whose functions may be realized in a single physical device, or distributed in multiple physical devices. The ASN (access service network) consists of all the radio components (shown within the network access provider domain), while the CSN (connectivity service network) consists of core network components (e.g., AAA server, user database and inter-working gateway). As shown, both visited and home NSPs (network service providers) have been accommodated in the network architecture via the R5 interface, while R2 defines a specific interface between a NSP and user terminal (subscriber or mobile station). While the ASP (application service provider) or the Internet is shown connected to the CSN, the interface between them is left open and is assumed implementation dependent.

Protocol Description

The WiMAX physical layer (OFDMA PHY) supports sub-channelization in both the DL and UL. The minimum frequency-time resource unit of sub-channelization is one slot, which is equal to 48 data tones (subcarriers).

There are two types of subcarrier permutations for sub-channelization—diversity and contiguous. The diversity permutation draws on subcarriers pseudo-randomly to form a sub-channel. It provides frequency diversity and inter-cell interference averaging. The contiguous permutation groups a block of contiguous subcarriers to form a sub-channel. The permutation scheme enables multi-user diversity by choosing the sub-channel with the best frequency response.

In general, diversity subcarrier permutations perform well in mobile applications while contiguous subcarrier permutations are well suited for fixed, portable, or low-mobility environments. These options enable the system designer to trade off mobility for throughput.

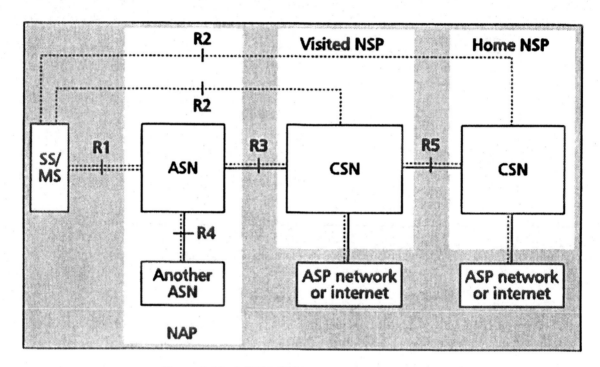

Figure 1-28: A WiMAX Network Reference Model

The MAC layer of IEEE 802.16 supports both bursty data traffic for bandwidth-intensive applications and lower-bandwidth continuous traffic for delay-sensitive (conversational) applications. The resources allocated to the user device by the MAC scheduler can vary from a single time slot to an entire frame, allowing a wide range of throughput variation. Additionally, the MAC scheduler can allocate resources ahead of sending each frame by including that information in the message before each frame.

QoS in the MAC layer of mobile WiMAX is implemented via service flow, as shown in Figure 1-29. Service flow is a unidirectional flow of packets (identified by SF ID) that contains specific QoS parameters associated with that WiMAX "connection" (a unidirectional logical link between the peer MACs at the mobile and base stations, identified by the term connection ID, or CID). The QoS parameters associated with the service flow include a traffic flow "classifier" and the "QoS category."

As shown, a protocol data unit (PDU, or basic message), whether in uplink (MS to BS) or downlink transmission, is sent to the classifier associated with the service flow where it is translated into a specific scheduler priority according to its QoS category. The scheduler then transmits that PDU based on its priority in relation to the resources available at that specific instant.

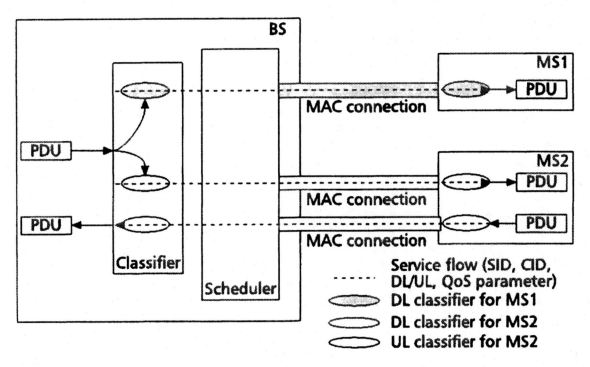

Figure 1-29: The MAC Layer QoS Model of Mobile WiMAX

Mobile WiMAX supports a wide range of data services and applications with varied QoS requirements. These are summarized in Table 1-5.

QoS Category	Proposed Applications	QoS Specifications
UGS Unsolicited grant service	VoIP and PTT	• Maximum sustained rate • Maximum latency tolerance • Jitter tolerance
rtPS Real-time packet service	Streaming audio or video	• Minimum reserved rate • Maximum sustained rate • Maximum latency tolerance • Traffic priority
ertPS Extended real-time packet service	Voice with activity detection (VoIP)	• Minimum reserved rate • Maximum sustained rate • Maximum latency tolerance • Jitter tolerance • Traffic priority
nrtPS Non-real-time packet service	File transfer protocol (FTP)	• Minimum reserved rate • Maximum sustained rate • Traffic priority
BE Best-effort service	Data transfer, Web browsing, etc.	• Maximum sustained rate • Traffic priority

Table 1-5: Mobile WiMAX Quality of Service Categories

Scheduling

The Mobile WiMAX MAC scheduling service provides optimized delivery of both delay-sensitive and delay-tolerant applications over broadband wireless. The main aspects of this service are:

- **Fast Data Scheduler**: The MAC scheduler at each BS allocates the available RF resources for bursty data traffic under time-sensitive channel conditions. Each PDU has well-defined QoS characteristics (associated with its SF) that allow the scheduler to determine OTA transmission priority.
- **Scheduling for DL and UL**: The scheduling service is provided in both DL and UL directions. Bandwidth request through a ranging channel, piggyback requests, and polling methods support UL bandwidth demands. The DL scheduler allocates resources based on the service flow priority.
- **Dynamic Resource Allocation**: The MAC scheduler allows frequency- and time-based resource allocation per-frame for both DL and UL. A MAP message delivers the allocation information at the start of each frame and manages the resources in response to variations in traffic and channel conditions. Furthermore, each resource allocation can consist of one slot or the entire frame.
- **QoS-Oriented Scheduling**: The scheduler provides QoS-based data transport per MAC connection as indicated earlier. The scheduler dynamically allocates or de-allocates RF resources in both the DL and UL based on the QoS parameters associated with each connection.
- **Frequency-Selective Scheduling**: The MAC scheduler can operate on various types of sub-channels. While using an AMC (adaptive modulation and coding)-based continuous permutation method, the sub-channels may vary in their attenuations. The frequency-selective scheduling can allocate the strongest sub-channel to a mobile user. Additionally, the method can increase system capacity with only a slight increase in CQI (channel quality indicator) overhead over the UL.

Mobility Management

Two important aspects of mobility management addressed by Mobile WiMAX—power consumption and seamless handoff—are introduced next:

- **Power Management:** Mobile WiMAX supports two modes—sleep and idle—within a *connection* state for efficient power consumption. Sleep mode is a connection state where the MS turns off its communication with the base station for a pre-negotiated time. Idle mode, on the other hand, presents a mechanism for the MS to be periodically available for DL transmission without re-registering with the base station. Idle mode allows the MS to be reachable in a highly mobile environment.
- **Handoff**: Three handoff methods are supported by the 802.16e standard—hard handoff (HHO), fast base station switching (FBSS), and macro-diversity handover (MDHO). Of these, HHO is mandatory while FBSS and MDHO are optional. The WiMAX Forum has developed several techniques for optimizing hard handoff within the framework of the 802.16e standard. These improvements have the goal of keeping layer 2 handoff delays to less than 50 ms:
  - In HHO, the MS must go through the entire registration process when it changes the base station.
  - In FBSS, the MS maintains continuous RF communication with a group of base stations called its active set. The MS is also assigned an "anchor" base station within this set that manages the UL/DL signaling and traffic bearer. Transition from one anchor base station to another depends on the signal strength measured by the MS on its CQI channel without an exchange of any explicit HO (handoff) messages. An important requirement of the FBSS is that the data destined for the MS is received by all base stations in its active set.

- In MDHO, the signaling and bearer communications over the RF resources are performed between the MS and all the base stations in its active set. This is done for both UL and DL transmissions.

Security

To provide security, Mobile WiMAX exploits advanced technologies. These include mutual device/user authentication, a flexible key-management protocol, strong traffic encryption, control and management plane message protection, and security protocol optimization for fast handoffs. Security is covered in Chapter 3.

1.8.3 Beyond 3G Vision and Long-Term Initiatives

1.8.3.1 ITU IMT-Advanced

ITU has defined systems beyond 3G, or "IMT-Advanced" (http://www.itu.int), by considering the needs of the evolving marketplace for future wireless communications, the required technologies, the spectrum, and other issues associated with the worldwide advance of IMT-2000 systems. These next, or fourth-generation, systems are evolving with a baseline defined by ITU-R Recommendation M.1645, "Framework and overall objectives of the future development of IMT-2000 and systems beyond IMT-2000" (approved in June 2003). This baseline was also considered at the World Radiocommunication Conference (WRC) in 2007 for identifying additional frequency spectrum for mobile and wireless communications.

IMT-Advanced goes beyond IMT-2000 to provide advanced mobile multimedia services with significant improvements in performance and user experience in both fixed and mobile networks. The key features of IMT-Advanced include:

- A high degree of commonality, flexibility, and efficiency
- Superior user experience and user-friendly applications and devices
- Inter-working of access technologies and global roaming
- Enhanced peak data rates to support advanced services and applications. Peak aggregate user data rates are envisioned of ~100 Mb/s for high user mobility and ~1 Gb/s for low user mobility (new nomadic/local-area wireless access).

Work on defining IMT-Advanced (in 2006-2007) focused on identifying technical requirements and evaluation criteria. These obviously depended on the availability of spectrum. WRC-07 (the ITU World Radiocommunication Conference 2007) tried to identify harmonized worldwide frequency bands for further development of IMT-2000 and IMT-Advanced. In total, close to 400 MHz of "new" spectrum for IMT was identified in "coverage" (lower) and "capacity" (higher) bands, taking regional considerations into account, and further study of sharing or harmonization requirements is needed. In addition, further studies of the greater long-term needs for spectrum were expected to be conducted and reported to WRC-11.

1.8.3.2 NGMN

The Next Generation Mobile Networks (NGMN) initiative consists of a group of mobile operators (known as members), together with other stakeholders in the mobile ecosystem (participants). It intends to provide a coherent vision for technology evolution beyond 3G and for the competitive delivery of broadband wireless services in the decade beyond 2010 (http://www.ngmn.org).

The vision of NGMN is to "provide a platform for innovation by moving towards one integrated network for the seamless introduction of mobile broadband services. The initial objective of the [NGMN] alliance is the commercial launch of a new experience in mobile broadband communications and to ensure a long and successful cycle of investment, innovation, and adoption of new and familiar services that would benefit all members of the mobile ecosystem."

The envisioned network architecture is packet-based and expected to provide a smooth migration of existing 2G and 3G networks towards an IP network with improved cost efficiency and performance. In addition to the technical challenges, NGMN addresses key issues related to intellectual property rights (IPR), inter-working, and aspects of operational services.

The details of the recommendations proposed by the NGMN initiative are captured in its White Paper, "Next Generation Mobile Networks Beyond HSPA & EVDO".

1.8.3.3 WWRF

The Wireless World Research Forum (WWRF) is a global organization founded in August 2001. With more than 140 members from five continents, it represents all sectors of the mobile communications industry and its research community. The forum's objective is to formulate a vision for the direction of future strategic wireless research by industry and academia, and to generate, identify, and promote research areas and technical trends.

The organizational structure of WWRF includes a number of Working Groups (WGs) and Special Interest Groups (SIGs):

- WG1: Human Perspective and Service Concepts
- WG2: Service Architecture
- WG3: Communication Architectures
- WG4: New Air Interfaces, Relay-based Systems, and Smart Antennas
- WG5: Short Range Radio Communication Systems
- WG6: Cognitive Wireless Networks and Systems

- SIG1: Spectrum Topics
- SIG2: Security and Trust
- SIG3: Self-Organization in Wireless World Systems
- SIG4: Convergence in Home and Enterprise Networks

The evolving research topics include:

- System concepts and high-level architectures
- Requirements of future mobile and wireless systems
- The role of scenarios for applications and services
- Applications and services
- Service categorization and service evolution
- Cooperative and ambient networks
- New approaches to self-organization in networking
- Meshing and multi-hop protocols for relay-based deployment

- Wideband channel measurement and modeling
- Duplexing, resource allocation, and inter-cell coordination
- Broadband frequency domain-based air interfaces
- Management and control architecture, scalability and stability of reconfigurable systems
- Cognitive radio and management of spectrum and radio resources in reconfigurable networks
- Business models and sustainable reconfigurability
- Requirements for future service platform architectures
- Determining spectrum efficiency and flexible spectrum use

The forum activities aim at creating a shared global vision to drive research and standardization for the future of wireless. In addition to influencing regional and national research programs, WWRF members contribute to the work done within the ITU, UMTS Forum, ETSI, 3GPP, 3GPP2, IETF, and other relevant bodies involved in commercial and standardization issues. As part of the effort to shape a common vision, WWRF has published a number of White Papers, addressing the technical areas of the Working Groups and Special Interest Groups. WWRF also produces the WWRF "Book of Visions."

The concept of the next-generation wireless system, and its design requirements and enabling technologies as envisioned by the WWRF members, are described in the WWRF System Concept document. The key requirements of such a system are the provision of "high data-rate transmissions and highly sophisticated services, comparable to those offered by wired networks and even going beyond. This will be achieved by disposing of a system architecture that is distributed, component-based and open. Such architecture will support context-awareness, personalization, and adaptation".

To achieve these requirements the WWRF system concept will rely on principles of ambient networking, cognition, dynamic spectrum management and self-organization, and a number of enabling technologies, such as smart antennas, relay-based systems, and cross-layer designs.

1.8.3.4 Other Initiatives

The Wireless World Initiative-New Radio (WINNER) project is a collaborative research initiative dedicated to addressing the challenge of 4G air-interface design and specifications. It was funded by the European Commission (2004-2007) and a large number of industrial and academic research centers in Europe, Asia, and Canada. The vision of WINNER 0 is to produce a ubiquitous next-generation radio system concept for providing wireless access for a wide range of services and applications, with adaptability to a comprehensive range of mobile communication scenarios and scalability in complexity. The aim is complete coverage, low deployment effort, and cost, and significant performance enhancements compared to legacy systems and their evolution.

WINNER has developed a system concept for a future mobile and wireless communication system by taking into account the ITU-R vision for IMT-Advanced. Compared to current and evolving mobile and wireless systems, the WINNER concept aims to significantly improve peak data rate, latency, mobile speed, spectrum efficiency, coverage, cost-per-bit and supported environments taking into account specified QoS requirements. The WINNER research work and findings have been extensively disseminated through publications and contributions to standardization bodies and the ITU.

Other Beyond 3G initiatives include the Next Generation Mobile Committee (NGMC) in Korea, the Future Technologies for Universal Radio Environment (FuTURE) national research project in China, and the neXt Generation (XG) communications Defense Advanced Research Projects Agency (DARPA) program in the United States.

In addition, standards bodies such as IEEE (e.g., 802.16m), 3GPP, and 3GPP2 continue to define a long-term roadmap through vision and system evolution committees, white papers, and new standardization work.

1.8.4 Technological Trends and Challenges

The radio interface for next-generation systems is expected to be packet-oriented and user-centric. Scalability and flexibility in design and deployment will be essential characteristics in order to adapt to varying conditions in the radio environment, the usage scenario, and network conditions.

The key attributes and goals of evolving wireless access technologies have been outlined in this chapter. In particular, further evolution of the systems to meet the principal objectives that have been set will be required to meet the long-term user, environment, and service requirements. Flexibility, efficiency, high-capacity, compatibility, inter-working, dynamic allocation, and adaptability will continue to be key requirements.

Personalization, adaptive and reconfigurable devices, ambient awareness, multi-modal "natural" user interaction in smart user spaces with "enhanced" reality, among others, are trends for vision, research, and innovation. Such technologies as cognitive radio, self-organizing networks and smart beamforming coupled with advanced technologies in other areas, such as sensing, robotics, and nano-engineering will enable this vision of the future of user-centric, context-aware, ubiquitous, and intention-based communications across all user spaces.

Technologies expected to play a decisive role in next-generation wireless-access system design include:

- The flexible air interface
- Hybrid duplexing, adaptive FDD/TDD
- Cross-layer optimization and integrated system design
- Relay-based enhancements, with networking through relay nodes and mesh evolution
- Interference management and inter-cell coordination, adaptive and reconfigurable inter-cell coordination.

Finally, note that identification of design objectives and requirements, and definition of technology innovations and solutions, will not succeed without adequate modeling and performance evaluation methods and systems. These concepts are not new but the environment in which wireless-access technologies operate evolves remarkably fast. The concepts demand new modeling and evaluation criteria in the presence of new user-interaction paradigms, dynamic user interfaces, and access mechanisms, traffic patterns, mixed services, interference sources, and time-varying communication parameters.

Chapter 2

Network and Service Architecture

2.1 Introduction

In a typical wireless network, a core network sits behind the radio access network, which is responsible largely for functions related to radio resource management. The core network, on the other hand, deals with switching/routing, mobility management, network quality of service (QoS), service provisioning, and service management. This chapter focuses on the core network; the radio access network has already been covered in Chapter 1.

There have been two major threads in the development of wireless networks. First came voice-centric cellular networks from the telephony world. The second thread, which rose to prominence quickly after the introduction of IEEE 802.11-based wireless LANs, encompasses wireless data networks from the data-communications world. The success of wireless LANs (WLANs) in the late 1990s has led to various concepts of integrating cellular and WLAN technology [Var2003], and it has been argued that rather than competing, WLAN and cellular technology can complement each other. In addition, fostering integration is the convergence of cellular and WLAN technology in 3G wireless standards. As the specifications of 3GPP and 3GPP2 evolve, they have both been moving towards more convergence, e.g., towards an all-IP network with the creation and development of the IP Multimedia Subsystem (IMS) [Won2003].

Exploring these developments will take up a significant portion of this chapter. We will also, however, discuss alternative wireless network architectures, service architectures, and teletraffic analysis.

2.2 Contents

This chapter addresses the following topics:

Traditional voice-centric cellular networks.

More data-centric, IP-centric network architectures, e.g., those used in wireless LAN networks.

The latest converged network architectures that include voice, video, and data services; IMS is an important example of such architecture.

Protocols and technologies for carrying voice and video over IP.

Alternative network architectures such as ad hoc and mesh networks.

This chapter also addresses the service architecture in wireless networks. By service, we mean a specific user application, which may be enabled by a particular technology platform or viewed as a customer solution. Examples include short message service and push-to-talk and location-based services. Service architecture is an arrangement in which different technology components, platforms, and service enablers are combined to compose and deliver services.

Last, but not least, teletraffic analysis, which is useful for network planning and capacity planning, is discussed briefly.

2.3   Circuit-Switched Cellular Network Architecture

First, consider the essential basic network elements in a traditional voice-centric cellular network. Since the terminology varies slightly from system to system, one system can serve as an example. For this purpose the GSM system, the most widely deployed 2nd-generation cellular system, is used. While the 1st-generation cellular systems, first deployed in the 1970s, were analog, 2nd-generation systems like GSM, IS-95 CDMA, and JDC are digital. Third-generation (3G) systems such as UMTS and CDMA2000, evolved from 2nd-generation systems, but have gradually evolved to more IP-based, converged networks (to be discussed in detail later in this chapter).

2.3.1 Basic Network Elements

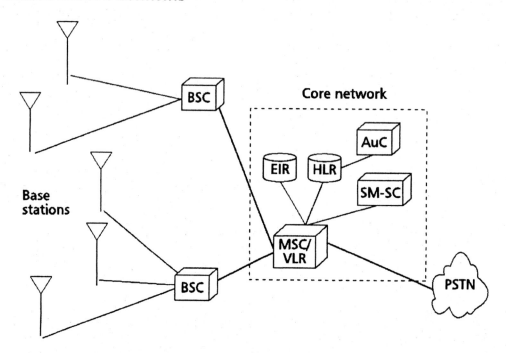

Figure 2-1: Network Architecture of a 2nd-Generation System

The essential elements of the core network of a 2nd-generation cellular system are shown in Figure 2-1. This chapter focuses on the elements in the core network, whereas, base station controllers (BSC) and base stations are involved in radio resource management, as discussed in Chapter 1. The core network contains a number of elements that together support telephony services, mobility management (roaming, etc.), security, short message service, and so on. Although GSM and its nomenclature have been used for illustration, other 2nd-generation cellular networks are similar.

The Mobile Switching Center (MSC) is like a normal switch in a telephone network but with additional functions to support mobility management. Thus, it understands standard Signaling System 7 (SS7) signaling between telephony switches, and supports the SS7 mobile application part (MAP). SS7 MAP uses the services of the SS7 transaction capabilities application part (TCAP) to communicate. The use of SS7 MAP will be illustrated shortly through a couple of examples. An MSC typically needs to be connected to other networks such as the public switched telephone network (PSTN), and suitable inter-working functions are needed for this. A service provider may have multiple MSCs in its network, especially if it is large. When an MSC plays a role as the first MSC to be reached by calls from other public land mobile networks (PLMNs), it takes on the gateway MSC role.

The home location register (HLR) and visitor location register (VLR) are databases that keep track of the location of mobile subscribers. However, they also contain other subscriber information such as subscription information and security data. The information for each mobile subscriber is typically stored in an HLR in the subscriber's home network. This would be the network of the provider with which the mobile subscriber has a subscription for mobile phone services. To speed up authentication when a subscriber is making a call? WHAT?, a VLR in the new network will temporarily store limited subscriber information in addition to security data.

To support roaming, the HLR and VLR together store enough information to allow roaming mobile stations to be located. When roaming, a mobile station registers its location in a process known as registration. Even when idle, a mobile station will send location updates as it moves between location areas (the concept of location areas will be introduced shortly), allowing the network to keep track of it. The location information is updated and tracked in the HLR and VLR as follows: the HLR points to the current network in which the mobile station is roaming, and the VLR contains more fine-grained information on its location. The VLR is often physically integrated with an MSC, and, hence, as in Figure 2-1, the MSC and VLR are combined into one network element, the MSC/VLR. It is, however, also possible to have separate MSCs and VLRs, e.g., when a service provider has multiple MSCs in its network, and they all share the same VLR.

The authentication center (AuC) is central to the security functions in the network

The short message service center (SM-SC) handles short message services, for the establishment of an end-to-end connection is not needed. Messages are first sent to an SM-SC, and then on to their destination. The SM-SC can store messages for a while when the destination mobile station is unreachable.

2.3.2 Call Flow Examples

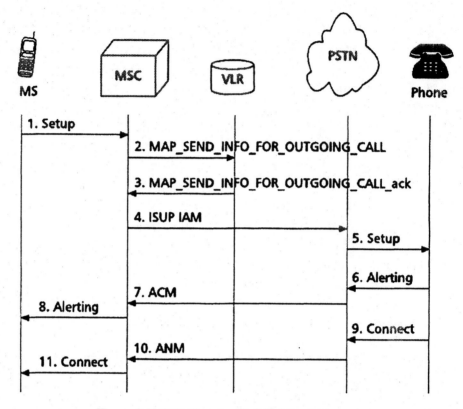

Figure 2-2: Call Flow for Call Origination

Next, consider how the elements of the core network work together. The first example is a call origination from a mobile station. The call flow is shown in Figure 2-2, which omits details such as the radio network signaling and radio resource allocation. Instead, these details are all lumped together under step 1, "Setup," which results in the MSC being requested to set up the call origination. SS 7 signaling is then used within the mobile network, for example, to query the VLR. Then, if the MS is allowed to originate calls and has enough of a balance (if prepaid), etc., an ISUP IAM message is sent to the PSTN. This is a standard initial address message in SS 7 to set up circuits. While the destination phone is being alerted (by ringing), an address complete message (ACM) is sent back to the MSC so the MS can be notified. When the phone is off-hook, an answer message (ANM) is sent.

In Figure 2-3, we see the more complicated case of call delivery to a roaming MS. Call delivery is more difficult than call origination because the MS needs to be located. The MS initiates call origination and the network receiving the signals from the MS uses the signal to determine the location of the MS. For call delivery to an MS, the signaling first goes to the MS's home network because the rest of the roaming network does not know where the MS is. As always when entering a mobile network, it enters through a gateway MSC. This MSC can query the HLR of the MS using MAP_SEND_ROUTING_ INFORMATION. As mentioned earlier, the HLR does not have complete location information of the MS so that the Gateway MSC (GMSC) cannot rely only on the HLR information to complete the call to the MS. Instead, the HLR knows which VLR to contact and it does so to obtain a crucial number, the roaming number. This roaming number is a temporary number that the GMSC can use to set up the next leg of the voice circuit towards the MS. The GMSC then forwards the ISUP IAM to the MSC in the roaming network serving the MS.

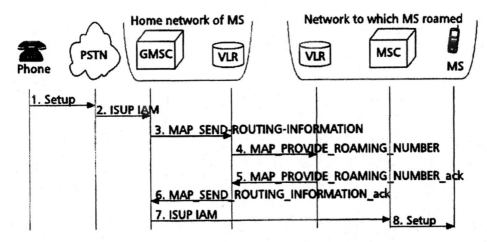

Figure 2-3: Call Flow for Call Delivery to a Roaming Mobile Station

Once the voice circuit connection set-up has reached all the way to the MSC serving the MS, one final challenge remains—to find the exact base station serving the MS. The MSC may not know precisely where the MS is if the MS is idle, in which case it only sends location updates to the network whenever the MS crosses location area boundaries. Typically, a location area would comprise multiple cells. The usual procedure for locating the MS more precisely within the location area is known as paging. Paging occurs when the MSC initiates a query to determine the location of the MS, i.e., paging messages for the MS would go out on all the base stations within the location area. Finally, the MS responds to the paging message at its current serving base station, and the MSC sets up the last leg of the voice circuit, completing the call delivery.

The size of a cellular system's location areas must be chosen carefully. However, the smaller the location area, the more often an idle MS must perform location updates. In the extreme of a location area with only one cell, the whole purpose of the location area is defeated—it is, after all, meant to reduce the frequency that the MS has to update the network. On the other hand, the larger the location area the more cells would be involved in paging during call delivery.

2.4   TCP/IP in Packet Switched Networks

TCP/IP is the predominant set of data network protocols in the world today. In the 7-layer Open Systems Interconnection (OSI) model of protocol layering, Transport Control Protocol (TCP) is a layer 4 (transport layer) protocol, whereas Internet Protocol (IP) is a layer 3 (network layer) protocol. The conjunction of TCP and IP in TCP/IP is due to the close historical relationship between TCP and IP, and even today, TCP is often used as the transport protocol over IP. However, as we will see in Section 3.5, TCP is not as suitable for traffic as VoIP, and alternatives like RTP are used.

TCP/IP was developed in the 1970s, based in part on 1960s research funded by the Advanced Research Projects Agency (ARPA). The name "internet" came about because one of the main goals of the network layer, and hence the Internet Protocol, was to interconnect networks of different types, using different layer 2 technologies. This concept of inter-networking proved very successful, and many universities and research institutions joined the Internet (a network using TCP/IP technology) in the 1980s. With the arrival of the World Wide Web (WWW) and the hypertext transfer protocol (HTTP) in the 1990s, the Internet started being used not only by researchers and students, but also by the general public.

Two main reasons that TCP/IP is important for wireless networks are:

- By the late 1990s, the IEEE 802.11 wireless LAN (WLAN) standard was introduced, and the use of WLAN technology exploded, especially for people with laptop PCs who appreciated the convenience of being able to connect to the Internet while on the go.
- Cellular systems have been evolving from voice-centric to more IP-centric, as will be seen in Section 2.6.

2.4.1 Basics of TCP/IP protocols

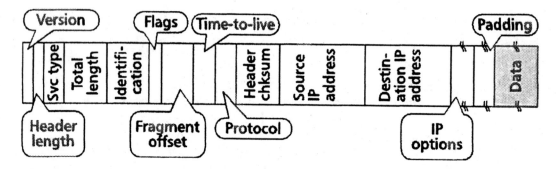

Figure 2-4: An IP Header

The IP header shown in Figure 2-4 contains all the information about a packet that a router needs to know in order to process it. More details on the IP header and the router processing of the IP header can be found in books on TCP/IP [Come2006].

User requirements have changed dramatically since TCP/IP was introduced over 30 years ago, as Internet usage expanded. The current version of IP is known as IPv4. A new version known as IPv6 [Deer1998] has been created, and is gradually being deployed around the world. Meanwhile, features continue to be added to IPv4, including mobility support (see Section 2.4.2).

IP hosts, the end points of an IP network, are often the source or destination of IP packets. IP routers, on the other hand, are intermediate nodes with multiple interfaces that forward incoming packets on outgoing interfaces based on the contents of the packet headers and routing tables. The contents and interpretation of routing tables can be found, for example, in [Come2006]. Several protocols have been developed to exchange routing information between routers so that the routing tables may have some degree of automated updating. These protocols include the router information protocol (RIP) [Malk1998], open shortest path first (OSPF) [Moy1998, Colt1999] and IS-IS [ISO2002].

2.4.2 IP and Mobility

Many of the contemporary uses for TCP/IP emerged after the protocol was initially designed. Thus, the use of TCP/IP for multimedia (with its different quality of service requirements), in wireless mobile networks, and so on, was not even imagined. With the rise of wireless networks, and especially wireless data networking, mobility support has become increasingly important.

Mobility should be distinguished from portability. Portability is the concept that a mobile wireless terminal can move and reconnect at each new attachment point with a new IP address. However, if portability is supported, certain applications will not work when the terminal moves if the applications expect the IP address to be unchanged. For example, for server applications, clients would typically perform a DNS look-up to find the server's IP address. With portability and IP address changes (of the server); the clients would have trouble finding the server. Furthermore, TCP sessions expect the IP address to remain the same. Changes in IP address will change the TCP checksum, for example.

Mobility support requires, therefore, the mobile terminal to keep its "permanent" IP address while moving, and at the same time still be reachable by other IP hosts. Mobile IP [Perk1996] solves this problem by allowing the mobile terminal to keep a permanent IP address as it moves. However, it relies on the existing IP routing fabric, and so for routing purposes, another IP address is required, known as a care-of address. The care-of address is another IP address routable to the mobile terminal's current location.

To allow IP packets addressed to the permanent IP address to be routed to the care-of address, Mobile IP introduces an agent, the home agent, in the mobile terminal's home network that intercepts packets addressed to the mobile terminal's permanent address, and forwards them to the care-of address. The packet forwarding is through a packet-encapsulation procedure where the original packet is encapsulated inside a new packet address to the care-of address, and tunneled to the care-of address. At the foreign network, another agent, the foreign agent, un-encapsulates the packet and delivers it to the mobile terminal. The home agent knows the current care-of address of the mobile terminal because of the Mobile IP registration procedure. In this procedure, the mobile terminal registers its care-of address with its home agent, which can then associate the two addresses. The operation of Mobile IP is illustrated in Figure 2-5. 5.

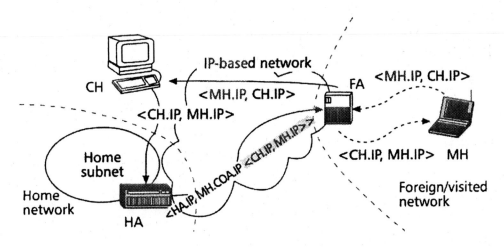

Figure 2–5: Basic Operations of Mobile IP

IPv6 also provides mobility support, which is sometimes called Mobile IPv6 [John2004]. It is similar in many ways to Mobile IP although Mobile IPv6 has some enhancements. These include:

- Support of route optimization.
- Support of fast handoffs.
- Reduced header overhead (using the IPv6 routing header).
- No need for foreign agents.

Furthermore, building on top of Mobile IPv6, not just host mobility is supported but network mobility as well. Network mobility (NEMO for short) is specified in RFC 3963 [Deep2005].

2.4.3 IP and GPRS

Mobile IP and Mobile IPv6 can provide mobility support in networks where the underlying link or bearer technology (that transports the IP packets) does not itself provide mobility support. A notable exception is GPRS. Two key networks elements in GPRS, serving GPRS support node (SGSN) and gateway GPRS support node (GGSN), act like routers but with the functionality to support the users' mobility. GPRS networks transport IP packets using the GPRS Tunneling Protocol (GTP) [3GPP29.060].

2.5   VoIP/SIP for IP Multimedia

Voice over Internet Protocol (VoIP) is one of the emerging technologies developed recently in the telecommunications industry. Compared to traditional circuit-switched voice service in the public switched telephone network (PSTN), VoIP promises a major reduction in capital and operational costs of the high-capacity packet-network infrastructure. With VoIP, it is also possible to create new revenue-generating services in a converged high-speed packet network. In particular, VoIP has been successful in wireline networks, with affordable commercial VoIP services available in the market. A set of new signaling and media-transport mechanisms has been defined. For example, Session Initiation Protocol (SIP) is used for call signaling control. Real-time Transport Protocol (RTP) is used for media delivery. Quality of Service (QoS) strategies and mechanisms are considered in the design of service platform and network infrastructure.

VoIP in wireless has drawn much interest recently because of the evolution to an all-IP architecture in converged wireless and wireline networks. Initial VoIP efforts were focused on wireless local area networks (WLANs) and fixed wireless since the achievable data rate is comparable to that of the wireline network. Recent advances in cellular high-speed packet networks have incorporated innovative techniques to improve spectrum efficiency and enable multimedia applications. Seamless integration and roaming of VoIP service across different wireless networks, i.e., WLAN, wireline VoIP, and cellular voice, will make VoIP not only a reality but also a mainstream application in wireless networks.

2.5.1 Protocols for VoIP Application

2.5.1.1 Session Initiation Protocol—SIP

The Session Initiation Protocol (SIP) [Schulzrinne-SIP] is an application-layer control (signaling) protocol for creating, modifying, and terminating sessions with one or more participants. These sessions include Internet telephone calls, multimedia distribution, conferences, presence, events notification, and instant messaging. SIP is specified in RFC 3261 [Ros2002] from the Internet Engineering Task Force (IETF) Multiparty Multimedia Session Control (MMUSIC) working group. SIP was adopted by both 3GPP and 3GPP2 standards as the main signaling protocol for multimedia over IP, and as a key element in the IMS architecture (see Section 2.6). SIP has been widely used as a call-control signaling protocol for VoIP applications in non-wireless networks.

SIP is a request-response and text-based protocol, very similar to HTTP (the protocol for WWW) and SMTP (the protocol for email). It is designed to be independent of the underlying transport protocol. SIP can run on top of TCP, UDP, or SCTP, which makes it very flexible and easy to implement. SIP consists of a set of messages that are used to set up and tear down a session. A simple example of a call setup and termination procedure using SIP (between user A and user B) is as follows, as shown in Figure 2-6:

1. The caller, user A, sends an INVITE message to initiate a call to the called SIP client, user B. The INVITE message consists of basic call setup information, such as the called and caller addresses, proxy information, and vocoder (voice coder) options.
2. The called party responds with a SIP TRYING message to acknowledge the call setup request.
3. The called party sends a RINGING message to the caller, where the appropriate ring back to the called can be generated.
4. A connection is established when the called party goes off-hook, and a media stream enabled. The called party generates an OK message to indicate the state change.
5. The caller connects the call and acknowledges the connection with an ACK message. The media stream for a two-way voice call is now established.
6. When the called party hangs up and the call is over, the called party sends a BYE message. The caller disconnects the call and acknowledges the termination with an OK message, and the call is terminated.

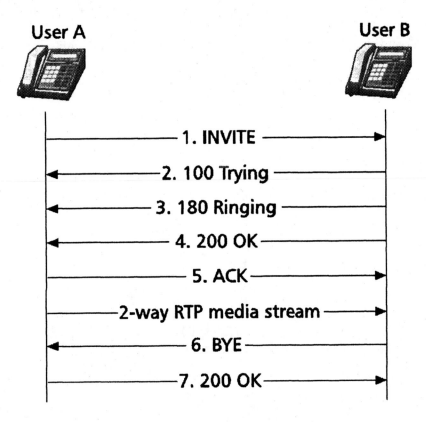

Figure 2-6: A Basic SIP IP Phone-to-IP Phone Call Flow

Meant for IP-based networks, SIP design intends to provide a set of call-processing functions and features similar to those in the current PSTN. SIP itself emphasizes on-call setup and signaling and does not define these PSTN-like features directly. It must work with other protocols and build these features in the network elements, such as proxy servers and gateways. An example of SIP network elements is the IMS architecture, described in detail in Section 2.6. An example of a generated SIP INVITE message is shown in Figure 2-7.

INVITE sip: UserB@ieee.org SIP/2.0
Via: SIP/2.0/UDP 200.101.200.005:5060
To: <sip:UserB@ieee.org>
From: User A <sip:UserA@1.1.1.1>
Call-ID: 1234567890
CSeq: 1 INVITE
Contact: <sip:UserA@1.1.1.1>

Figure 2–7: An Example of a SIP INVITE Message Format

2.5.1.2 Real-Time Transport Protocol—RTP

The Real-time Transport Protocol (RTP) [Schulzrinne-RTP] is a standard protocol for transporting multimedia applications over the Internet. It was developed by the Audio-Video Transport working group of the IETF and first published in RFC1889 in 1996, and then in RFC 3550 in 2003 [Sch2003]. It defines a packet format for delivering audio and video packets before passing them to the transport layer. RTP can be used for transporting common voice formats, such as PCM, ITU standard voice codec G.711 and others, GSM codec, audio format such as MPEG audio (MP1, MP2, MP3), and video formats such as MPEG, H.263, H.264, etc. It can also be used for transporting proprietary audio and video formats. RTP has been widely implemented for streaming multimedia applications.

RTP usually runs on top of UDP because real-time applications have stringent delay requirements. The sender takes a media packet or data chunk and encapsulates it into an RTP packet. It then encapsulates the RTP packet into a UDP packet and passes it to the IP layer. The receiver extracts the RTP packet from the UDP packet, then extracts the media packet or data chunk from the RTP packet and passes it to the media layer for decoding and play-out.

As an example, consider the use of RTP to transport a G.711 voice stream. The G.711 encoder generates a compressed voice stream at 64 kb/s (an 8-kHz sample frequency × 8 bits per sample). Assume that the application layer collects the encoded data every 20 ms, or equivalently, 160 bytes of data, and passes it to the RTP layer. The sender encapsulates the 160 bytes into an RTP packet with a header. The RTP header is usually 12 bytes, including payload type (7 bits), sequence number (16 bits), time stamp (32 bits), synchronization source identifier (SSRC, 32 bits), and miscellaneous fields (9 bits), as shown in Figure 2-8. The 172-byte RTP packet (a 160-byte payload plus 12-byte header) is passed to the UDP socket interface and sent to the receiver. The receiver extracts the RTP packet from the UDP socket interface. The application layer extracts the voice payload from the RTP packet and uses the header information to decode and play out the voice stream.

Misc.	Payload type	Sequence number	Time stamp	SSRC
9 bits	7 bits	16 bits	32 bits	32 bits

Figure 2–8: RTP Header Format

RTP is used purely for media transport over the Internet and can be viewed as a sub-layer of the transport layer. It provides a mechanism for easy interoperability between networked multimedia applications. Note, however, that RTP does not provide any QoS guarantees of timely delivery of the applications. The RTP encapsulation is visible only at the end points, and is invisible to routers or switches along the path. RTP is often used in conjunction with other protocols and QoS mechanisms to provide VoIP service.

2.5.1.3 Robust Header Compression (RoHC) Protocol

VoIP packets are often carried over RTP/UDP/IP, with an uncompressed header of about 40 bytes per voice frame. The voice codec used in cellular systems often has a very low bit rate, from 4 kb/s to 12.2 kb/s, to save bandwidth in the air interface. Consider a 22-byte voice frame (equivalent to an 8 kb/s full-rate voice codec). The RTP/UDP/IP header overhead is more than 65% (~40/62). The radio application protocol stack adds additional overhead. Without any header compression, the overhead becomes dominant and has a significant impact on the air-interface capacity. A header-compression scheme therefore becomes an essential component in providing VoIP in wireless.

Robust header compression (ROHC) [Bor2001] is a standard specification of a highly robust and efficient header-compression scheme. It was developed by the ROHC working group of IETF, and published in RFC 3095 in 2001. It specifies a framework supporting four profiles, i.e., uncompressed, RTP/UDP/IP, UDP/IP, and ESP/IP. ROHC offers the ability to run over different types of links by operating in different modes based on the availability of a feedback channel and how much feedback information can be provided. With ROHC compression of RTP/UDP/IP headers, the average header size is between 1 and 4 bytes, and within a voice stream most of the packets can carry the header information in only a single byte. The protocol stack with header compression is shown in Figure 2-9.

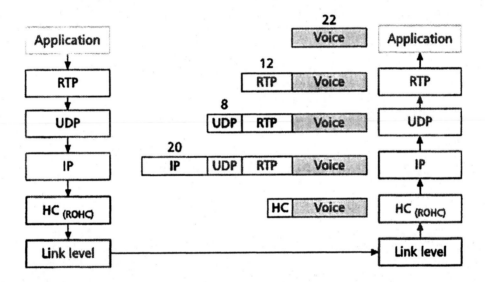

Figure 2-9: A Protocol Stack with Header Compression

Header compression relies on exploiting the redundancy in the headers. RTP, UDP, and IP headers have significant redundancy in both intra- and inter-packets of a media stream. An intra-packet has identical or deducible information. For example, the source and destination address in the IP header and the source and destination port in the UDP header do not change for the same media stream; they only need to be transmitted once at the media stream's start. The payload type in the RTP header only needs to be sent once as well.

An inter-packet has incremental differences in the same stream. For example, the time stamp in the RTP header increases by a fixed amount with every increase in the sequence number. The time stamp need not be sent to the ROHC decoder at every talk spurt along with the sequence number. It must be sent only at the beginning of a new talk spurt after a silence compression.

Note that ROHC is not used as an end-to-end compression scheme since routers need the full IP header information for routing purpose. It is often used in the radio-access network and improves the air-interface capacity for VoIP applications.

2.5.2 *Quality of Service (QoS) for VoIP in Wireless*

One of the most important requirements for Internet telephony is to provide Quality of Service (QoS) guarantees, with service classes defined for different applications. For example, a common set of QoS definitions include conversational, streaming, interactive, and background classes. QoS metrics usually include delay, delay jitter, bandwidth, and reliability, which vary for different applications. Most stringent is the delay required for conversational voice applications, with the end-to-end packet delay less than 200 ms. Packet loss rate should be less than 1%. The mean opinion score (MOS) is often used to provide a numerical measure of human speech. The score is generated by using subjective tests to obtain a quantitative indicator of speech quality. There are five scores ranked as 1 (bad), 2 (poor), 3 (fair), 4 (good) and 5 (excellent). Acceptable voice quality requires a MOS score of at least 3.5. A specification of MOS is part of the Perceptual Evaluation of Speech Quality (PESQ), which is standardized by ITU. PESQ is better suited to measure the quality of VoIP than an earlier framework, PSQM.

Providing end-to-end QoS guarantees for VoIP is extremely challenging because the wireless infrastructure is heterogeneous and operated by different service providers. The common practice is to provide QoS mechanisms at each network segment so that the overall end-to-end QoS can eventually be met. At the radio access network (RAN) side, the QoS design includes flow classification, a packet scheduler over the air interface, backhaul buffer management and control, call admission and overload control, and header compression for improved efficiency. At the core network side, the QoS design includes support of DiffServ for priority routing and handling, MPLS for ensured bandwidth and service guarantee, and IMS session control for multimedia applications. The end-to-end QoS mechanism includes user authentication and negotiation, de-jitter play-out buffer design, and end-to-end delay improvement.

2.6   Packet-Switched Mobile Networks and IMS

2.6.1 *Background*

With the addition of over two billion users in 25 years, cellular telephony has been a phenomenal success. The goal of first-generation analog cellular systems was to provide voice telephony to mobile users even though voiceband data was feasible to a certain extent. Second-generation digital cellular systems introduced short message service (SMS) and circuit-switched data. SMS became very popular and enjoyed widespread acceptance especially among GSM users in Europe and the Far East. Even after data capability was introduced in second-generation systems, voice telephony continued to be the major application.

Circuit-switched connection is very inefficient for data services when it comes to scarce radio bandwidth resources. The bursty nature of data traffic means that the channel bandwidth is too narrow when the user has data to send and, when there is no data, the channel is idle most of the time. The advantage of packet switching is well known when a large number of users with bursty data traffic share a common medium. GSM introduced a packet-overlay network in the core mobile network, as well as a packet protocol over-the-air interface, to make better use of radio spectrum and network resources. Known as General Packet Radio Service (GPRS), it was a key step in the evolution towards packet-switched core networks and all-IP mobile networks. (GPRS was discussed earlier.)

2.6.2 Evolution to Packet-Switched Core Networks and All-IP Mobile Networks

As third-generation cellular systems evolved, supporting higher speed data over the air interface became a key requirement. The ITU, which defined the high-level requirements for third-generation wireless, set forth the data rates to be supported in mobile, pedestrian, and fixed environments. In the meantime, the emerging Internet was changing the way people accessed information, and its widespread popularity made wireless data services even more appealing.

Telecommunications has seen two major trends lately. The growth in voice services has slowed, resulting in voice revenues that are flat or even decreasing. Data traffic is growing rapidly and, with it, a major shift towards packet-switched networks. In the early days of telephony, when voice dominated traffic, data services could be supported in the voice-centric network. With the growth in data traffic, network providers were left with the choice of either maintaining two separate networks—one for voice and one for data—or finding ways to support voice traffic in data networks. Since maintaining two different networks is not economically attractive, evolution towards a single packet-switched core network began in earnest.

The trend towards packet networks in cellular systems started with the introduction of the Internet protocol (IP) in the core network, as shown in Figure 2-10. Traditional mobile switching centers (MSCs) are decomposed into MSC servers, which provide the call-control functions of the MSC, and into media gateways (MGWs), which provide conversion between circuit-switched voice and packet voice. Interworking with legacy circuit-switched systems is achieved with the help of additional signaling gateways that convert signaling over IP transport to signaling over SS7 networks. Even though the core network has all the signaling and voice traffic transported over an IP network, call control for voice services continues to use existing protocols, and voice over-the-air interfaces use circuit-switched connections.

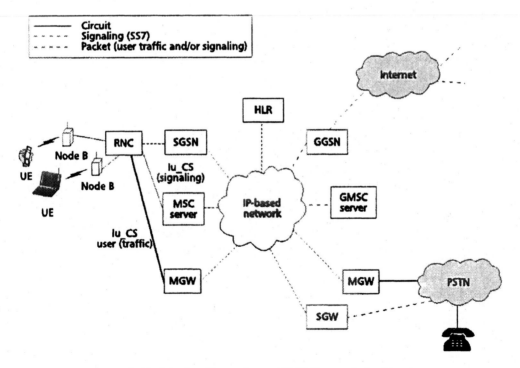

Figure 2-10: Moving Towards an All-IP Network Architecture

2.6.3 The Need for IMS, and Its Requirements

As cellular systems evolved to 3G, voice services remained as the major application. Even though 2.5G and 3G cellular technologies provided higher bandwidth access, packet-switched services did not take off as expected. The "killer application" for data continued to elude everyone, even though native connection to the Internet is available through wireless access, and the access bandwidth increased with new air-interface technologies. Service providers realized that providing higher bandwidth access alone was insufficient and that rich data services were required. The voice-centric cellular infrastructure had limitations and could not take full advantage of IP-based services. It became apparent that a new approach was needed to stimulate IP-based services. IP multimedia subsystem (IMS) [Won2005] emerged as the new approach that could eventually stimulate IP services in wireless networks.

The vision of IMS is to offer an open architecture with a common control/session layer that promises the best of the fixed, cellular, and Internet worlds. IMS also attempts to replace the vertical stovepipe model where each technology has its own horizontal layered architecture of core network, call control, supplementary services, and applications. In this architecture, different access technologies (DSL, cable, cdma2000, UMTS, WiMAX, and WLAN) can share a common IP transport layer, common control/session layer, and a common pool of applications. By merging wired and wireless networks, IMS [Won2003] brings (a) the broadband capability of fixed networks, (b) the convenience of mobility from the cellular networks and (c) rich user experience of new and ever-expanding applications from the Internet. The objectives and requirements for IMS include [3GPP TS23.228]:

- Enabling operators to offer IP multimedia services.
- Enabling a mechanism for negotiating QoS.
- Supporting interworking with circuit-switched (CS) networks (PSTN, cellular) and the Internet.
- Allowing roaming
- Providing home network control.
- Allowing rapid service creation.
- Supporting third-party development of IP-based services.
- Providing access independence.
- Relying on IETF-approved standards.

With access bandwidth increasing as new air-interface technologies are introduced, and native connection to the Internet available through wireless access, the need for IMS and the role of the operators have been questioned in many quarters. The argument is that subscribers be allowed to try third-party applications and use third-party service providers for IP multimedia services. Though there may be some benefit to this approach, it may not be in the best interests of both operators and subscribers.

From an operator's point of view, offering bandwidth as a commodity and letting users choose their own Internet services will shut out the operators from lucrative new service offerings. Indiscriminate use of Internet applications over cellular networks could also affect the integrity of the cellular network and quality of service. However, operators cannot anticipate all the applications users may want or be able to develop these applications, because this is not their area of expertise. In addition, such an approach could stifle the applications development seen in the Internet domain. IMS offered a new approach where the operators could provide the session control, while developing some applications themselves. At the same time, third parties would be allowed access to some network services and interfaces where they could develop the majority of applications.

2.6.4 IMS Architecture and Network Elements

The IMS architecture, shown in Figure 2-11, is a collection of functions linked by standard interfaces. The architecture includes the IMS terminal, commonly referred to as user equipment (UE), IP connectivity access networks (IP-CAN), one or more SIP servers, collectively known as call state control functions (CSCFs), one or more databases, application servers, and nodes to interwork with legacy circuit-switched networks and media resource functions. The elements in the cloud are the nodes of the IMS core, which interface with other IMS networks. The transport plane is a core IP network of routers. In IMS, signaling and media planes are completely separate. The only entity that handles both is the UE.

The UE can be a VoIP, data, or multimedia terminal. The UE attaches to the IP-CAN. Examples of IP-CAN are GPRS, W-LAN, cdma2000 packet network, WiMAX, and cable or DSL access.

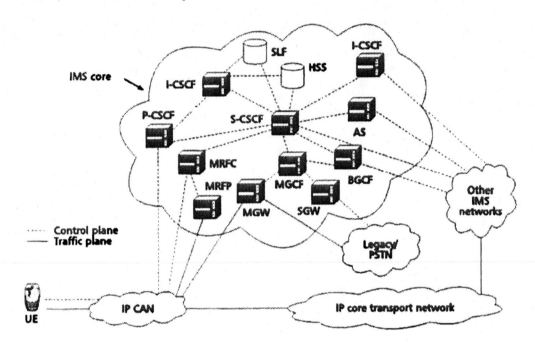

Figure 2–11: The IMS Architecture Links a Collection of Functions with Standard interfaces

Home subscriber server (HSS) is a central database for user-related information. It is an evolution of the home location register (HLR) from the cellular world and contains user information such as subscription, location, security, and user profile to support multimedia sessions. IMS allows more than one HSS in a network. Subscriber location function (SLF) is a database that maps a user address to a given HSS. Networks with only a single HSS do not need an SLF.

Call state control function (CSCF) is a key function in IMS. They are SIP servers and process SIP signaling in IMS. There are three types of CSCFs:

- The proxy CSCF (P-CSCF), the first point of contact between the UE and IMS network, acts as an outbound/in-bound SIP proxy server. Each IMS network includes a number of P-CSCFs, each serving a number of IMS terminals. The P-CSCF is usually located in the visiting network when the mobile is roaming.
- The S-CSCF is the central node in the IMS. In addition to SIP server functions, the S-CSCF also performs session control and enforces the policy of the network operator. The S-CSCF also acts as a SIP registrar binding the IP address of the terminal to the public address. The S-CSCF is always located in the home network, which may include a number of S-CSCFs.

- The I-CSCF is a SIP proxy at the edge of an administrative domain. Its address is listed in the DNS and allows CSCFs to route the SIP messages from one IMS network to another via an I-CSCF.

The signaling gateway (SGW), the media gateway control function (MGCF), and the media gateway (MGW) comprise the PSTN gateway and allow IMS terminals to make and receive calls from other CS networks. The MGW converts media from packets to CS and vice versa. The MGCF converts between SIP messages and messages understood by the CS networks. In addition, the MGCF controls the MGW. The SGW performs the lower-layer protocol conversion between IP and CS domains for the messages between the MGCF and a switch in the CS domain. The breakout gateway control function (BGCF) determines where in this network, or in some other IMS network, the PSTN breakout should occur, and decides which MGCF to involve.

The media resource function (MRF) provides the ability to play announcements, mix media streams (for conference services), and transcodes between different codecs. The MRF includes a signaling component (MRFC) and the processing component (MRFP).

An application server (AS) is a SIP node that hosts and executes IMS services. There are three types of application servers. A SIP application server hosts and executes IMS services on SIP. IMS-specific services are likely to be developed using SIP ASs. To use existing applications developed for OSA and for GSM, a service capability server (SCS) is included to convert the protocols so that these ASs look to the IMS networks like SIP ASs.

2.6.5 IMS Procedures

IMS is designed to use IETF protocols as much as possible. The most widely used protocol is the session initiation protocol (SIP). SIP is used between the UE and the P-CSCF, between CSCFs, between the S-CSCF and the AS, between the S-CSCF and the BGCF, between the BGCF and the MGCF, and between the S-CSCF and the MRFC. Diameter is another protocol used in IMS. Diameter is used between the CSCFs and databases like the HSS and the SLF, as well as between the AS and the databases. Diameter is used for authentication, authorization, and accounting (AAA) in IMS. H.248 is used between the MGCF and the media gateway, as well as between the MRFC and the MRFP. The MGCF uses stream control transmission protocol (SCTP) between the MGCF and the SGW. Real-time protocol (RTP) is used for end-to-end delivery of real-time data. RTP provides time stamps, among other things, and allows the receiver to play out the media at a proper pace.

Like mobile phones, IMS terminals need to register before they receive services from the network. It is during the initial registration that the network assigns an S-CSCF for the subscriber that establishes a path between the UE and the S-CSCF for IMS call control. Before the terminal can register, it needs to gain access to the IP-CAN and acquire an IP address. Depending on the IP-CAN, the IP address may be obtained as part of getting IP connectivity (in GPRS) or obtained separately (in DHCP). The terminal needs to discover the P-CSCF and get the IP address of the P-CSCF. In GPRS, the PDP context-activation response provides the P-CSCF address; otherwise, DHCP and DNS are used. Registration involves many sub-procedures. It binds the public user identity to a contact address, home network authenticates the subscriber, the subscriber authenticates the network, and the UE and the P-CSCF establish a security association for further signaling. Once registration is completed, the terminal stays in an idle mode until an active session needs to be established.

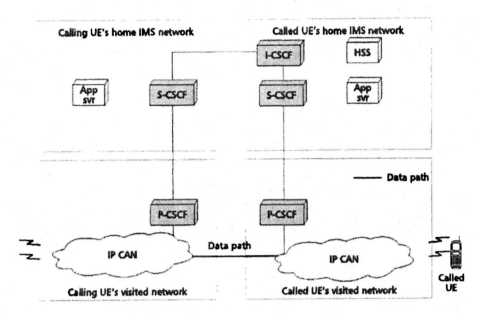

Figure 2-12: Call Establishment in IMS

Figure 2–12 shows the steps involved in the most general case of a session establishment where both calling and called party are roaming outside their home networks. The left side of the figure shows the visiting and home networks for the caller, while the right side shows the visiting and home networks for the called user.

Session establishment starts when the calling UE sends a SIP INVITE message to the P-CSCF in the visiting network. The INVITE includes the SDP describing the parameters of the session and the public user identity of the called user. The P-CSCF processes the INVITE request and forwards the SIP INVITE to the S-CSCF in the user's home network. The S-CSCF performs session control and uses the filter criteria to invoke applications from application servers. An example of the use of AS could be to check the call-barring list for this user. There could potentially be a chaining of application invocations.

Using the public user identity of the called user, the S-CSCF resolves the SIP URI and the IP address of the I-CSCF in the home network of the called user and forwards the INVITE. The I-CSCF queries the HSS and gets the address of the S-CSCF serving this user. The S-CSCF performs service control and invokes the AS service, if required. The S-CSCF then forwards the INVITE to the P-CSCF in the visited network. Session establishment continues with the negotiation of session parameters and preconditions and provisional acknowledgments.

When the called user answers the session, a 200 OK is returned to mark the successful completion of the session setup in response to the original INVITE message. The calling UE returns an ACK as the final response. When the session setup is complete, the data path is established between the calling user's IP-CAN and the called user's IP-CAN. Both IP CANs in the illustration are in the visited networks, and the data path may traverse intermediate IP networks. Signaling and media planes in IMS are completely separate and signaling in IMS always traverses the home network. This is unlike a traditional cellular system where, for a call made from the visited network, the signaling need not traverse the home network.

For IMS to be viable, it needs to interwork with legacy circuit-switched (CS) networks like PSTN and existing cellular systems. When a call to a phone in the CS network is made by an IMS user, the SIP INVITE arrives at the S-CSCF of the calling user. The address analysis will indicate that the call is destined for a user on the CS network and the S-CSCF forwards the INVITE to the BGCF, which determines to which MGCF the INVITE should be forwarded. The MGCF converts SIP signaling to a signaling protocol that the CS network can understand and forwards the signaling to the switch on the CS network via the SGW signaling gateway. The MGCF also involves an MGW, which performs media conversion between packet voice and circuit-switched voice. In the reverse direction, for a call originating from the CS network and destined for a user on the IMS, the CS network forwards the signaling to the MGCF via the SGW. The MGCF then converts the signaling to SIP INVITE and forwards the INVITE to the S-CSCF for further signaling.

Interworking between the IMS and CS network requires translation between SIP signaling and SS7 signaling, and transmission of SS7 messages between the MGCF and the SGW. IMS resides in the IP domain, and the transport between MGCF and the SGW is over IP networks. SS7 messages must therefore be transported over IP networks. To transmit these messages reliably requires acknowledged and error-free transmissions. While UDP cannot ensure reliable acknowledged transmissions, TCP is too restrictive which may affect the performance (e.g., delay requirements). To address this, IETF formed a working group known as SIGTRAN to define specifications for reliable transport of CS network protocols (e.g., SS7 and ISDN) over IP networks. It uses stream control transmission protocol (SCTP) to provide transparent transport of message-based signaling protocols. The scope of SCTP includes definition of encapsulation methods, end-to-end protocol mechanisms, and IP capabilities that support the functional and performance requirements for signaling. SCTP provides the following functions [RFC 2960]:

- Acknowledged error-free non-duplicated transfer of user data.
- Data fragmentation to conform to discovered path MTU size and sequenced delivery of user messages within multiple streams, with an option for order-of-arrival delivery of individual user messages.
- Optional bundling of multiple user messages into a single SCTP packet.
- Network-level fault tolerance through support of multi-homing at either or both ends of an association.

As with CS interworking for voice calls, IMS needs to interwork with the Internet. This means that the S-CSCF may need to exchange SIP messages with a SIP server located outside IMS. These external clients may not support one or more of the SIP extensions required for IMS. In this case, the SIP user agents within IMS may fall back to basic SIP. Operators may also decide to restrict session initiation with external SIP clients that do not support extensions.

When IMS was originally designed, IPv6 was expected to be widely available by the time IMS was deployed. SIP and associated protocols have also had problems with the network address translation (NAT) traversals required in an IPv4 network. For this reason, 3GPP (originally) decided to use only IPv6. However, although IMS is ready to be deployed, IPv6 has not taken off, and IPv4 and NATs are becoming ubiquitous. Thus, 3GPP allowed IPv4 in early IMS, with dual-stack implementations (IPv4 and IPv6) allowed in IMS terminals and nodes. Two elements, IMS-application-level gateway (ALG) for SIP interworking and transition gateway (TrGW) for RTP interworking have been added to IMS architecture.

2.6.6 *IMS as a Platform for Convergence*

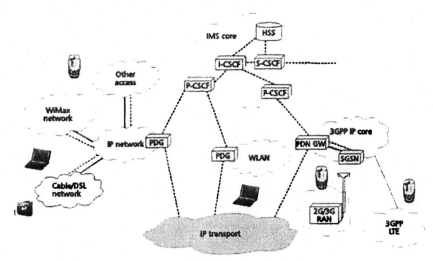

Figure 2-13: IMS as a Platform for FMC

Though IMS was conceived, and is being standardized in 3GPP, it is becoming a platform for fixed mobile convergence (FMC). There are multiple access networks today, including TDMA in GSM/GPRS access, UTRAN, cdma2000 access in 3GPP2, WLAN, cable, DSL, and WiMAX. More access networks are expected in the future. These will include LTE (long-term evolution) networks, as defined in 3GPP, and different variations of 4G wireless networks espoused by academia and industry.

Instead of defining all the layers as in a vertical silo network, it is beneficial to have a common service and control layer that can support multiple access networks; both wireless and fixed networks could evolve while providing backward compatibility. With the access network agnostic control and service layer, IMS provides an ideal platform for multiple access networks and for FMC. An example of how IMS can provide the platform for FMC is shown in Figure 2-13. A packet data gateway (PDG) or a packet data network gateway (PDN GW) at the edge of the access network will allow IMS to serve as the common control and session layers. Though not shown in the figure, IMS will provide open APIs to the application layer to make the same applications available to users independently of the access network.

IMS relies on an underlying IP core for signaling and media transport. Several efforts are underway in 3GPP to realize a common IP core network that provides IP-based network control and IP transport across, and within, multiple access systems. The requirements and objectives of this effort, defined in the all-IP network (AIPN), include;

- a seamless user experience for all services within and across the various access systems.
- the ability to transfer sessions between terminals.
- a high level of security.
- QoS.
- high network performance.
- advanced application services.
- the ability to select the appropriate access system based on a range of criteria.
- interworking with existing fixed and wireless networks.
- support for a wide variety of terminals.
- efficient handling and routing of IP traffic.

The system architecture evolution (SAE) is another ongoing development in 3GPP for an IP packet-optimized network infrastructure to support the growing demands for IP traffic in terms of both increased data rate and reduced latency. The SAE aims to make optimum use of the increased bandwidth available in access networks and to support multiple access technologies, including existing and evolving 3GPP access networks and non-3GPP access networks. Non-3GPP access networks include wireless LANs (WLANs), 3GPP2 access networks, and WiMAX. A key objective of SAE is service continuity and terminal mobility between different access networks. Mobility must be supported at different levels including within the evolved 3GPP and between 3GPP and non-3GPP systems. This mobility needs to provide for the seamless operation of both real-time and non-real-time services. The SAE effort does not directly impact or influence IMS procedures. However, IMS can take advantage of the core IP network to provide a stable IP address for SIP signaling.

2.6.7 *Summary*

IMS provides an open architecture, with a common control and session layer, to offer the best of the fixed, cellular, and Internet worlds. The 3GPP initiated IMS, but a number of other standards bodies soon embraced it. IMS introduces a number of new network elements. The call state control function (CSCF) is the most central node in IMS. Service control in IMS is always in the home network. This differs from the philosophy of traditional cellular networks that puts service control in the visiting network. IMS allows support for applications from both home and third-party networks. IMS protocols are based on IETF protocols; of these, SIP and Diameter are the most dominant. IMS allows interworking with CS networks and the Internet. IMS users can make and receive calls from CS networks users and also establish sessions with Internet users and applications. Work is ongoing to support seamless handoff between IMS and circuit-switched domains. With access-agnostic control and session layers, IMS is an ideal platform for fixed mobile convergence. A number of operators have IMS trials underway, and some have already started early deployments. We anticipate deployments and a healthy growth of IMS subscribers in the coming years.

2.7   Alternative Network Architectures—Mesh Networks

The wireless mesh network (WMN) is a new concept revolutionizing the way future broadband Internet access could be provided to customers. The concept involves mesh clients (MCs). Basically, the idea is to place wireless routers (known as mesh routers, or MRs) on top of tall buildings where they form a static ad hoc network among themselves. Each MR serves neighborhood MCs. A few MRs known as Internet gateways (IGWs) are also connected to the Internet and provide Internet access for the rest of the MRs. When the MRs and IGWs are owned and operated by different individuals, the configuration is known as a community-based network. When MRs are controlled by a single service provider, the arrangement is called fully managed. An intermediate solution known as a semi-managed WMN is possible when MRs are controlled by a few entities. An example of such a generic WMN is shown in Figure 2-14.

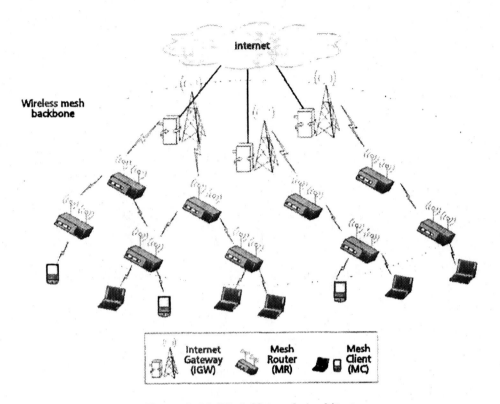

Figure 2–14: Mesh Network Architecture

An important issue in implementing a MESH network is to make sure that MCs have sufficient wireless access coverage and bandwidth. In order to deploy an adequate number of MRs, it is important to know the number of users and their total bandwidth requirements. It is also useful to know if MRs can be placed anywhere in a given geographic area or if there are limits on their location. Moreover, it is useful to know the busy period of each MC and operating schedule of each MR, which are indications of the instantaneous traffic load and the number of MRs needed. MRs can be located where AC supply is readily available and, hence, the power constraints present in a typical mobile ad hoc network (MANET) [Cor2006] do not apply.

Another important issue involves WMN positioning. The topology of WMNs can be formed on an ad hoc basis, with the connectivity pattern depending on the relative distance between two adjacent MRs. Unlike a MANET, MRs are not mobile, the topology remains static unless, due to business decisions, some MRs are taken out of commission or turned on at a later date. The location of the MR and the IGW is one of several critical factors that determine WMN performance [Gup2000]. If the MRs and the IGWs are randomly situated, as in a MANET, a WMN can encounter the several problems: (1) unbalanced load distribution, (2) uncontrolled interference, and (3) unreliable architecture.

The WMN positioning depends on the physical configuration of the IGW and MRs, including their location and the number of their interfaces. Their configuration is subject to constraints involving geography, the maximum number of channels in the network, and traffic demand. Of particular concern is where the IGWs should be placed to minimize their number while still satisfying the MR Internet throughput demand. Important IGW selection algorithms include cluster-based IGW selection [Bej2004], OPEN/CLOSE heuristic IGW selection [Pra2006], and tree-based IGW selection [He2007].

The efficient placement of MRs is a challenging issue because of many practical constraints and contradictory requirements, including cost, link capacity, wireless interference, and varying traffic demands. To minimize costs while meeting traffic demand, a given region must be covered with a minimum number of MRs and interfaces. Thus, goals for MR positioning include providing enough network capacity, avoiding congestion incurred by balancing traffic, maximizing network capacity with a limited number of MRs and their configured interfaces, and insuring that the network is fault tolerant. An elementary exploration of MR placement is presented in [Wan2007]. The authors discuss MR placement on the condition that MRs can only be placed in pre-decided candidate positions, while considering coverage, connectivity, and traffic demand constraints. They propose a two-phase heuristic algorithm to find the close-to-optimal MR placement.

Note that traffic in a WMN is generated by MCs at the MRs, and moves between MRs and IGWs and not between the MRs themselves. In fact, MRs needing access to the Internet either download the packets from an IGW or forward them to reach IGW. Each MR, besides serving MCs in its own neighborhood, ought to forward packets toward an IGW from MRs further away, and the volume of traffic near each IGW is the highest. The volume of data increases closer to the IGW with the maximum data rate dictated by the cumulative bandwidth of the IGWs. Thus, responsibility for routing must be undertaken by each MR. Note that because packets must be queued and served in each MR, packets generated by MRs near an IGW tend to achieve higher bandwidth than MRs further away. Therefore, utmost care must be taken to process packets based on the number of MRs already traversed [Nan2007].

Even after the architecture of a WMN has been designed, many issues remain regarding the utilization of resources. Decisions are needed regarding the number channels to be allocated to MRs and the number of radio interfaces for each MR. This can affect the simultaneous utilization of multiple channels at one or more MRs, reduce interference between them, and increase the availability of bandwidth at different MRs. Accordingly, it is desirable to have more and more wireless radios as one gets closer to the IGW; such a design can manage the increasing concentration of traffic near the IGWs. Note that interference between neighboring channels should be avoided. Merely having an adequate number of radios is not enough because it takes a finite time to change the status of an interface circuit to on or off, and to modify the radio frequency.

Multiple channels are available in current MAC protocols. For example, 12 orthogonal channels are available with IEEE 802.11a and three with IEEE 802.11b. An MR could tune its radio interface to a channel not being used by neighboring transmitting MRs to avoid interference and improve throughput. For higher throughput, an MR equipped with a single radio interface can dynamically switch the frequency among multiple channels. The efficiency depends on the delay required for an interface to switch from one channel to another, which is not negligible.

Another important factor is the channel-switching synchronization for a pair of transmitting MRs whose time information could be from either an external source, such as the Global Position System (GPS), or an internal source, such as beacon signals from neighboring MRs. For effectively exploiting multiple channels, each MR should be equipped with multiple interfaces that work simultaneously on different channels. Each interface of an MR can be fixed at a channel that is not used in the interference region. If the channels are statically bounded to multiple interfaces, the number of channels utilized by each MR is constrained by the number of interfaces, which may force the removal of some links because a pair of MRs, within the transmission region, may not have a common channel. Hence, while maximizing throughput, the assignment of channels to interfaces needs to ensure the network's connectivity. Appropriate assignment of multiple channels to interfaces can eliminate a substantial amount of interference and enhance the network's performance.

It is also necessary to determine the optimal path from a given MR to an IGW. One can find the shortest paths from each MR to all IGWs and select the IGW closest to the MR, although the shortest path might be not be the best choice because MRs are known for incorporating multi-rate links. The shortest design may be fine if the traffic is originating only from that particular MR, but, with the MRs free to serve their MCs, the volume of traffic passing through each MR can fluctuate significantly and each MR must forward the packets towards the IGW. Therefore, availability of bandwidth along a given path not only depends on its own traffic, but also on the traffic generated by other MRs and the paths selected to reach the IGWs. In selecting an optimal path for a given flow, it is important to have a global picture of the volume of flows through the network. Information is needed about the multiple paths from each MR to one or more MRs. Such dynamic selection of the paths and channels is important because the WMN topology may also change with time as some of the MRs are turned off and on.

Another important issue that arises in the design of a large-scale WMN is the fairness of the MCs, regardless of their distance to the IGW. It is well known that the IEEE 802.11 network suffers from degradation of throughput in a multi-hop path because of interference, hidden terminal problems, and so on. Careful attention should be paid to ensure that traffic from nearby MCs does not dominate the buffers of MRs near the IGW, and that sufficient buffer space is guaranteed to be available to distant flows. This requires efficient queuing-management and traffic-splitting schemes [Nan2007a].

As discussed earlier, besides serving their own MCs, all MRs must cooperatively forward packets towards the IGWs. However, some MRs might act selfishly so as to fully utilize the available bandwidth and provide higher Internet throughput for its MCs. This is a serious problem and such "free-riding" behavior is unacceptable in a community-based WMN network. Owing to the hierarchical architecture of WMN, MRs further away from the IGW are inherently dependent on the MRs near the IGW to forward their packets. A selfish MR near the IGW can thus cause serious performance degradation. Packets dropped by the selfish MR have already consumed significant network resources as they are forwarded by other intermediate MRs.

Thus, such free-riders ought to be penalized by some kind of detection mechanism. Detecting selfishness in a multi-channel WMN becomes a challenging problem. Some traditional reputation-based approaches that depend for detection on promiscuous listening might be applicable due to the assignment of non-overlapping channels between adjacent MRs. Unfortunately, due to the static nature of WMNs, a credit-based approach that assigns virtual currency to cooperative MRs largely fails since MRs in the periphery of the network fail to earn sufficient credits and are handicapped. Traffic-monitoring techniques have been devised to detect and identify such misbehavior [San2006], and then isolate such MRs by not serving their associated packets.

The 802.11s standard was developed by the IEEE 802.11 working group for wireless mesh networks. Depending on wireless distribution, the architecture of 802.11s supports broadcast, multicast, and unicast. Its main objective is to perform routing at the link layer. The interconnected wireless devices in 802.11s mesh networks are called mesh points (MPs). MPs may be static or mobile, but they have packet-relaying capability and can connect to one another automatically to form a mesh topology. Some MPs in the network have wired connection to the Internet and act as IGWs. Therefore, unlike a traditional 802.11 WLAN, a mesh network provides Internet connectivity over multiple wireless hops. A MP is able to discover any existing WMNs within the transmission region and can associate with them. An MP is also able to initiate a new network if no network is detected. The whole discovery process is auto-configured and does not need any user intervention. MPs in the network can exchange packets over multi-hop wireless links, selecting the path dynamically. When an MP needs to send a packet, it performs layer 2 MAC-based routing, which is different from layer 3 routing using IP addresses. Furthermore, the routing metrics of the link layer depend on the quality of the wireless links because the layer observes channel conditions and finds the route(s) satisfying the QoS requirements. Therefore, 802.11s is capable of guaranteed performance with respect to such parameters as throughput, packet loss, delay, jitter, and so on.

2.8   Alternative Network Architectures—Mobile Ad Hoc Networks

The Mobile ad hoc network (MANET) is one of the newly emerging wireless network architectures. It is composed of wireless user devices that deliver data packets via multi-hop packet forwarding. Like wireless mesh networks (WMNs), MANET differs from conventional wireless networks (e.g., cellular and WLAN networks), which are infrastructure-based and have the wireless link just at the last hop. Both WMN and MANET are multi-hop wireless networks, but WMN nodes are fixed while MANET nodes can move around. As shown in Figure 2-15, source node S transmits data packets through a multi-hop route to destination node D. En route, MANET nodes act as intermediate relay nodes to forward data packets for other users

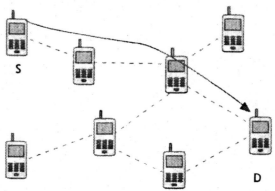

Figure 2-15: Routing in a Mobile Ad Hoc Network

Routing, which discovers the multi-hop packet-forwarding paths, is a major issue in MANET, and the user's mobility poses challenges in the routing protocol design. For example, the routing protocol needs to discover new routes while keeping track of the validity of old ones. Based on routing discovery and updating mechanisms, the MANET routing protocol can be categorized as either proactive or reactive.

Ad hoc on-demand distance vector routing (AODV) [Per2003] and dynamic source routing (DSR) [Joh2001] are two popular reactive routing protocols. These protocols, also known as on-demand routing protocols, discover multi-hop relay routes only when needed (on-demand).

In AODV and DSR routing protocols, the basic mechanism of finding valid routes is control message flooding. When an AODV source node S would like to send a data packet to a destination node D, S first checks its routing table to see if a valid route to D exists. If a valid route exists, S just sends the data packet to D using that route; otherwise, S will initiate the route-discovery process. In this route discovery process, S floods the route request (RREQ) message to all nodes in the network. S first broadcasts the RREQ message to its neighboring nodes. When intermediate network nodes receive a RREQ message, they re-broadcast the message to their neighboring nodes. When the destination node D receives the RREQ, which indicates S is looking for a route from S to D, the destination node D replies with a route reply (RREP) message toward S. During the RREP transmission from D to S, AODV intermediate nodes store the routing information for a valid route between S and D. After S receives the RREP message, the valid route toward to D is recorded, and S then starts sending data packets to D.

Similar to the AODV operations, DSR initiates the route-discovery process if there is no valid route entry for sending data packets. The flooding of route request (RREQ) and replying of route reply (RREP) is similar to that of AODV. The main difference between AODV and DSR is that DSR is a source-routing protocol that carries routing information in packet headers. During the route-discovery phase, DSR nodes store the hop-by-hop routing information in RREQ and RREP routing messages. During the data-transmission phase, the routing information is carried in the data packets' header.

Proactive routing protocols operate similarly to the Internet routing protocol. They maintain valid routes by periodically updating routing information. Destination sequenced distance vector (DSDV) [Per1994] and optimized link state routing (OLSR) [Cla2003] are two popular proactive MANET routing protocols. OLSR is based on a link-state routing mechanism in which network nodes distribute their neighboring connectivity to the whole network to compute the valid routes. DSDV is based on a distance-vector routing mechanism in which network nodes distribute their entire routing tables to their neighboring nodes to compute the valid routes.

In DSDV, each node periodically forwards the routing table to its neighbors with a sequence number that indicates the freshness of the routing information. When a wireless node forwards its routing table, the sequence number is increased and appended to the routing table. The received-neighbor routing tables are stored with the local routing information, with their sequence number attached. This number is used to guarantee loop-free routing. In addition, DSDV is designed to overcome fast topology change in MANET by sending an immediate route advertisement when there are significant changes in the routing table. This feature enables fast routing updates for topology changes in MANET.

In conventional link-state routing protocols, all nodes periodically broadcast the status of their neighboring links. In OLSR, the link-state routing is optimized for MANET by reducing the broadcast of link-state information. Unlike wireline networks, wireless transmission is broadcast in nature. A transmitted message is received by all nodes in the neighborhood of the sender. As a result, not all MANET nodes need to re-broadcast messages to achieve network-wide flooding. Only subsets of OLSR nodes, called multipoint relays, are selected to broadcast link state information. The multipoint relay nodes are selected so that the link-state information flooding covers the whole network while reducing redundant transmissions.

There are also hybrid routing protocols that combine the mutual advantages of proactive and reactive routing. For example, zone routing protocol (ZRP) [Pea1999] applies proactive routing within the routing zone closer to the ZRP wireless node, while applying reactive routing outside the routing zone. Moreover, some MANET routing protocols combine with location information in routing operations [Ko2000]. These types of location-aided, or geographical, routing operate effectively when the wireless network topology is highly correlated with the geographic topology.

MANET technology has great potential for a wide variety of applications. The technology has been adopted by the U.S. Army for military applications, and can also be applied to vehicular wireless networks. The vehicular ad hoc network (VANET) is an extension of MANET suitable for high-speed and predictable mobility scenarios. Wireless mesh networks (Section 2.7) could also be viewed as an extension of MANET. In wireless mesh networks, some infrastructure nodes are added as dedicated relays to assist multihop forwarding; hence, user device forwarding is reduced to achieve better network scalability. Wireless mesh networks and MANET technology have been applied to municipal and community wireless-access applications.

2.9   Service Enabler Evolution

2.9.1 Background

A broad range of services has been provided by a variety of wireless access technologies over the years. The key mobile communication service has been voice call, and cellular systems have traditionally been circuit-switched and optimized for voice. Mobile data services have, however, grown significantly. As much as 15% to 30% of mobile business, in a variety of global markets, may be attributed to non-voice services. The evolution of packet/IP-based networks enables efficient development, control, integration, and delivery of IP multimedia services. At the same time, a converged service framework increasingly allows the creation and delivery of all services in a converged and common environment, allowing orchestration of services and distribution of common functions.

2.9.2 Mobile Services

Mobile data services have been available for years, particularly with messaging, but also with browsing, downloads, and location-based services in second-generation networks. The growth of mobile networks, user terminal functionality, and user experience, have fostered rich communications, mobile email, high-speed access and business services, and browsing and content services, all of which are growing significantly. This is an integral part of serving an increasingly mobile society that is evolving to demand further personalized and community services and applications for such new paradigms as work, health, entertainment, information, advertising, and user-generated content on the go.

This evolution is not just about speed but also experience and value. The evolution will succeed in its response to the need for a user-centric, and not a platform-centric, service space that is intention-based. A service object may be dynamic and virtual. In space, it may include one or more enablers choreographed to serve the intention; in time, it can adapt to the changing parameters, needs, or communication space. However, without capacity growth and cost efficiency, the exponential growth in wireless traffic cannot be accommodated. These are two key design objectives for systems beyond 3G, as discussed earlier.

The term "service" has been used to refer to a particular user application, which may be enabled by a particular technology platform or viewed as a customer solution. It has also had other uses as network functions, XML messages, and others. In this section, service enablers are referred to as particular technology standards or platforms that typically work end-to-end to enable a particular service (or application). A service enabler may be able to provide more than one user application, particularly as services become more application-based, context-aware, and personalized or user-centric. Alternatively, multiple enablers may be leveraged to create a service. For example, a rich messaging service may be created by use of multimedia messaging, presence, location, and possibly group (XML list) enablers.

Short message service (SMS) has been a leading and superbly successful mobile data service. Typically, it includes text with a limited number of characters, mostly to communicate short messages between mobile terminals. However, variety does exist in its applications across systems, in its broadcast capability, in concatenated schemes, and in broad use of SMS messaging within networks for other services. SMS supports both mobile originated (MO) and mobile terminated (MT) operations. It has a client-server architecture in which an SMS Center manages the operation of store and forward, re-attempts, and other administrative tasks. Because messaging, like conversational services, is expected to be seamless across users, inter-operator solutions exist, possibly through gateways

The main global standardization body that specifies the mobile service enablers is the Open Mobile Alliance (OMA), which works in close coordination with other standards bodies and industry forums. The standards, which make use of IETF protocols, try to be bearer-independent, though they may accommodate bearer-specific objects and elements where needed. They include such service enablers as messaging and rich communication, browsing, content delivery and management, broadcast and multicast, location, presence, device and rights management, mobile commerce, and mobile advertising.

Proprietary services have been implemented broadly for one business need or another, including time-to-market. There is a great deal of effort, however, to establish interworking across systems and service providers.

Messaging paradigms have evolved and become richer, allowing near-real-time user interaction and leveraging other enhancers, such as user mobility information. These paradigms include mobile multimedia messaging and mobile instant-messaging enablers, in addition to push-to-talk over cellular. The general trend is towards a seamless user experience (e.g., converged messaging).

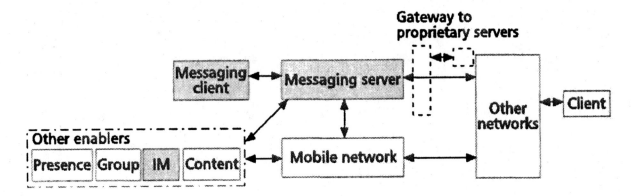

Figure 2–16: A Simplified View of a Messaging Enabler

Push-to-talk schemes are typically half-duplex communication formats in which one user has the floor at a time. It allows communication with one user or a group of other users simply and nearly instantaneously. The standardized push-to-talk over cellular (POC-OMA) enabler can provide a modular architecture with a number of defined interfaces, leveraging other services or functions. Figure 2-16 shows a functional diagram of a messaging enabler and figure 2-17 shows the interactions between entities.

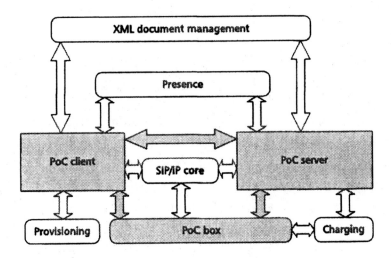

Figure 2-17: Interaction of PoC Functional Entities with External Entities

Location-based services are intuitively necessary in the context of mobility. They include emergency, navigation, and tracking services, in addition to other enhanced services such as location-based commerce, charging, advertising, and content delivery. The location function can use different technologies, ranging from cell/sector information, to combined use of cellular network (e.g., location of receiving base stations) and signal (e.g., round-trip-delay or triangulation) information, and GPS or assisted GPS, among others. Standardization work has allowed use of different position-determining mechanisms while defining how location is triggered and delivered. Figure 2-18 shows a high-level block diagram for a location enabler, for emergency and commercial services. Privacy loop is an important component getting the user's authorization, except when assumed for emergency services.

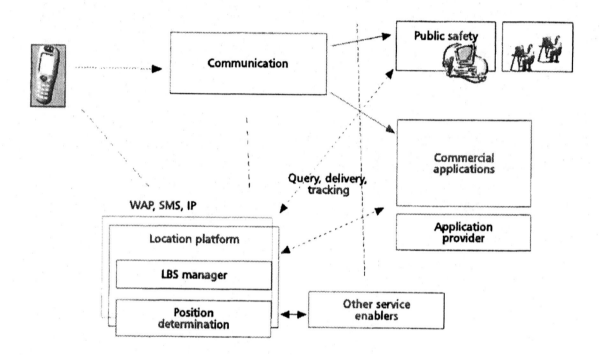

Figure 2-18: A Simplified Block Diagram of Location-Based Services

Other service enablers, as noted earlier, have been introduced, defined, or are emerging:

- Content management and its dynamic and personalized delivery.
- Broadcast and multicast service-enabler definitions, leveraging external distribution systems.
- Mobile commerce, both radio-based (RFID) and optical based (e.g., 2D bar codes).
- User-generated content and social networking.
- Device management for the life cycle of the user terminal, with such over-the-air features as activation, provisioning, diagnostics, firmware updates, software distribution, and others.

The reader is encouraged to explore sources in this context, including the industry forums developing standards (notably www.openmobilealliance.com) or applications definitions, profiles, certifications, interoperability testing, and guidelines.

2.10 Service Framework

2.10.1 Background

Technologies and architectures for the enabling of services have grown remarkably, yet have not reached their full potential. The evolution is about:

- Rich services and rich experience with increasing cost efficiency.
- Need for open and secure interfaces to the application space.
- Need for a converged and common service framework to orchestrate services and functions to serve the user.

There is a need for open, secure, and modular integration of services, and for granularity and abstraction to be able to compose and port services, functions, and support systems. In other words, a common service framework will allow orchestrated and efficient service creation, discovery, access, and delivery.

There has been a flurry of activity within the industry to define open interfaces and service architectures, and to outline a common service framework. Note, however, that there is no single approach to this, either in space (best practice, implementation-specific, co-existence with IP core, coexistence with the so-called Web 2.0 and simplified connectivity for mass social networking) or in time (phased evolution).

2.10.2 Open Service Framework

Typically, converged service frameworks have a number of common attributes, no matter the difference in architecture, protocols, implementation, or best practice. Service enablers, though different in their nature, are developed to "expose" themselves, to register on a logical framework, or to be discovered. There are open interfaces, defined as external bodies such as application providers, business customers, virtual operators, or others, allowing binding using SIP, web services or other protocols. In addition, a common framework should be able to distribute, port, or re-use functions, and to package/compose them given the abstraction inherent in the exposure.

Openness is about the use of universal (Internet) protocols that are network-independent. It refers to a combination of IT and telecom capabilities, allowing access to networks and services by internal and external entities through the use of known application programming interfaces (APIs). Furthermore, openness allows distribution and sharing of functions. The service framework, however, has measures of security and rules of engagement (or binding) based on policies.

Different service enablers that register onto a framework may have their own specifications and protocols. Therefore, a level of abstraction may be specified to "expose," discover, or use these functions or enablers in an open-access environment. The following are examples of actions for which an open, secure, efficient, and common service framework are needed:

- To distribute content broadly and simply.
- To choreograph and re-use functional elements such as business-support systems, policy, charging, and so on.
- To re-use horizontal enablers such as location and presence.
- To integrate multiple service elements (e.g., location, streaming, presence, m-commerce, messaging) to create a composed intention-based service (e.g., a user asking about a movie).

This framework distributes functions, integrates multiple services to form a composite serving the user's intention, and allows access by internal and (trusted or authorized) external entities. An example of such a framework is shown in Figure 2-19.

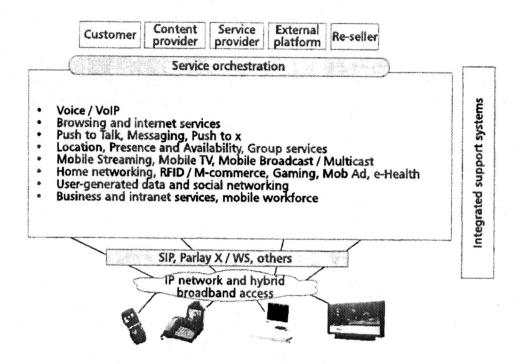

Figure 2-19: An End-to-End Service-Enabling Framework

Technologies such as OSA/Parlay (http://www.parlay.org), with interfaces such as Parlay X/Web Services and concepts such as service-oriented architecture (SOA), service creation environment (SCE), or service delivery platform (SDP) relate to this space and are recommended to readers for further exploration. For example, an SOA solution may be used in an operational environment to evolve from an environment with a large number of different operational and business-support systems to a shared and "open" one, where portability, re-usability, and integration are possible in a flexible and cost-effective manner. Similarly, new services may be created efficiently and with richness, or dynamically accessed, composed and delivered, leveraging multiple enablers and functions, internal or external, which would otherwise be in silos or hard to discover, leverage, re-use, or integrate.

In addition, IP multimedia subsystem (IMS) standards allow use of a Parlay gateway when the application server is not SIP-based. Furthermore, IMS is also concerned with common session control, common client environment, and the concept of brokering. An end-to-end system integration includes the service architecture and the IP core. This is no surprise, as the IP core should ultimately and increasingly be viewed as an IP multimedia service framework (higher layer), while increasingly enabling cooperative sub-networks and seamless mobility (lower layer). The reader is further directed to 3GPP and 3GPP2 standards and such concepts as IMS and system architecture evolution, in addition to joint work with the Parlay group.

2.10.3 Service Environment Evolution—Trends

As hinted in earlier sections, the evolution of nomadic and high-mobility service environments is increasingly about enabling rich communication, IP services, and multimedia content management and delivery. A common and converged user-centric service framework will allow the mixing of elements/functions, or multiple service enablers, to create and deliver an intention-based service. The so-called horizontal enablers such as location, presence, quality-of-service, policy, and group management have a central role to serve on their own or to enhance other enablers.

As service-enabling environments aim to become more user-centric and intention-based, such concepts as personalization and dynamic delivery of content (e.g., DCD at OMA) become important.). The definition of service itself is dynamic and fluid, as a platform-centric approach increasingly becomes user-centric and context-aware, serving such new or growing paradigms as a mobile workforce, social networking, e-health, information and entertainment on the go, nomadic access to education, and communities.

2.11 Fundamentals of Traffic Engineering

Wireless network operators strive to provide high-quality service in a cost-effective manner. Service quality can be defined at many levels, including but not limited to the quality of the voice encoding and error correction. . At the level of wireless channel availability, the quality of service can be defined by availability metrics, two of the most important of which are:

- Blocking probability: the probability that a new call (either incoming to a subscriber or outgoing from the subscriber) is *blocked,* i.e., it cannot be allocated the necessary network resources to proceed.
- Dropping probability: the probability that an existing call is *dropped* during a handoff, i.e., the necessary resources cannot be allocated in the new cell for the handoff to complete successfully.

A call that is blocked is referred to as a blocked call and a call that is dropped is referred to as a dropped call (an event also called forced termination). The blocking and dropping probabilities are easily expressed as blocking rate and dropping rate, respectively.

Wireless network operators face a tradeoff between providing high-quality service, and providing it cost-effectively. On the one hand, the best quality service can be interpreted as meaning 100% availability, where network resources (e.g., wireless channels) are always available for customers to use. With 100% availability, there are no blocked calls and no dropped calls. Thus, high-quality service requires the provisioning of an over-abundance of network resources. On the other hand, cost considerations would tend to limit the network resources to be provisioned. The tradeoff can be visualized in Figure 2–20.

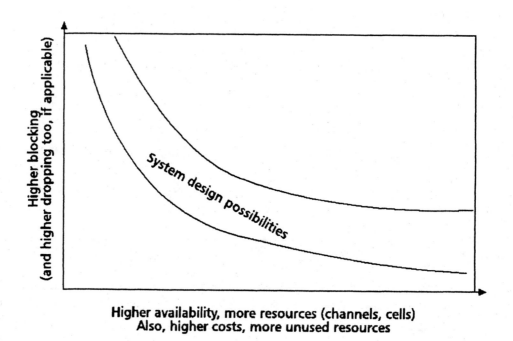

Figure 2-20: Tradeoff Between Quality and Cost

Teletraffic analysis deals with network availability such as call-blocking rate, call-dropping rate, channel-occupancy rate, and so on. Such analysis is of great interest to wireless service providers because they can use it to set the parameters of their network given the limits of radio resources and network resources available. For example, they may want to minimize the call-dropping rate. Sometimes, this may come at the expense of call-blocking rate (which is the rate that new calls are blocked), since subscribers are more likely to be annoyed by the dropping of existing calls due to the blocking of handoffs, than by occasional failure to begin a new call.

With the universal trend towards convergence, integrated wireless networks (where subscribers may move back and forth between cellular and WiFi or other wireless technologies) are becoming increasingly popular. Thus, the networks are becoming more complicated and it is becoming more difficult to do network planning (setting parameters like the number of channels per base station). Therefore, teletraffic analysis is becoming increasingly essential. Moreover, to differentiate between cellular and WLAN cells, multi-layer, hierarchical modeling methods are typically used, which can be computationally very complex.

2.11.1 Teletraffic Analysis for Wireless Systems

There is an extensive body of literature on teletraffic analysis for wireless cellular systems. Because of handoffs between cells, the standard models used for decades in telephony had to be modified for use in wireless networks. Furthermore, with different traffic types, networks with different types of cells (e.g., mixed microcells and macrocells), and the introduction of features like guard channels, the problem can be quite challenging. In the great majority of the papers and books, Markov models are used because of the ease in applying standard queuing theory analysis and results to the problem. Even if the actual traffic characteristics may not be exactly Poissonian, it is hoped that it is close enough so that the analytical or simulation results obtained by using a Poisson process model are still meaningful.

It must be noted that the Poisson process assumption has been challenged by some researchers (especially the handoff traffic, rather than new call traffic [Chle1995]), as a result of which newer models have been proposed that are supposedly more accurate. An example of such a model is the proposed use of a Hyper-Erlang distribution for the inter-handoff time [Fang 1999], rather than an exponential distribution. Closed-form analytical expressions (like the Erlang B and Erlang C formulas) are often not derivable (even when using Poisson assumptions) because modeling of wireless networks is far more complex than modeling standard telephony networks. Thus, in general, computer simulations are used to obtain teletraffic performance of wireless systems.

One of the main thrusts of research is how to differentiate the treatment of handoff traffic and new call traffic. Handoff calls are often given higher priority than new calls, since subscribers would find it more annoying if existing calls are dropped when handoff fails, than for new calls to be blocked [Sidi1996]. Several methods for such prioritization have been proposed. One method uses guard channels [Oh1992, Oliv1999], and, instead of a fixed number of guard channels, performance may be improved (at the expense of complexity) by using a variable number of guard channels [Choi2000]. Another method to reduce dropped-call probability is known as channel borrowing [Katz1996]. Channels are borrowed from neighboring cells to accommodate handoff calls.

The complexity of the standard queuing system model grows exponentially as the number of states increases, i.e., as the number of channels and cells increases. Approaches to simplifying the teletraffic analysis problem to something more computable have generally focused on simplifications to the model itself, or on ways to simplify the computations for a given model.

Of the attempts to simplify the model itself, making many assumptions related to homogeneity and loose coupling, some researchers have attempted to obtain useful information about a whole network by using a one-cell model. The idea is that if the coupling between cells is "loose enough," it suffices to pick a single cell in the network and examine the teletraffic dynamics of that cell alone to have a good idea what is happening in the network. The cell can be arbitrarily chosen, based on homogeneity arguments. The incoming/outgoing traffic from/to neighboring cells may be approximated as Poisson processes [Ekic2001]. It may be shown that such an approach allows simplification to a well-known M/M/m/m Markov model [Ekic2001, Orti1997], whose closed-form solution is easily available from standard queuing theory. However, the accuracy of single-cell models is very questionable for more sophisticated wireless networks that employ channel borrowing, dynamic guard channels, etc. Furthermore, it is not useful for multi-layer networks.

One class of simplification approach is the use of sparse matrix algorithms. Because of loose coupling, the matrix built from the global balance equations is a sparse matrix. Sparse matrix algorithms have been investigated for use in solving packet network models [Kauf1981]. More recently, sparse matrix algorithms have also been used to study hierarchical wireless networks [Zhou 1998]. Nevertheless, the savings in computation time are limited.

2.12 Key References

D. Comer, Internetworking With TCP/IP Volume 1: Principles Protocols, and Architecture, 5th edition, Prentice Hall, 2006.

M. Mouly and M.B. Pautet, The GSM System for Mobile Communications, Telecom Publishing, June 1992.

K.D. Wong, Wireless Internet Telecommunications, Artech House 2005.

Abbas Jamalipour, The Wireless Mobile Internet: Architectures, Protocols and Services, John Wiley and Sons, 2003

Chapter 3

Network Management and Security

3.1 Introduction

In this chapter, we turn our attention to aspects of the management of wireless networks. We start our consideration of these topics from the perspective of two distinct operational process models:

1. The *Information Technology Infrastructure Library* (ITIL), which is maintained by the Office of Government Commerce (OGC) of the United Kingdom, and
2. The Enhanced Telecom Operations Map (eTOM), which is maintained by the TeleManagement Forum (TMF).

We will use these two perspectives to examine in some detail the underpinning technologies and protocols with specific attention to:

1. Fault and performance monitoring, and
2. Security

The intent of this chapter is to highlight the important network-management capabilities pertinent to wireless communications.

3.2 Contents

This chapter addresses the following topics:

The Information Technology Infrastructure Library
The Enhanced Telecom Operations Map
The Simple Network Management Protocol
Security Requirements

3.3 The Information Technology Infrastructure Library

The government of the United Kingdom describes the *Information Technology Infrastructure Library* (ITIL) as a public domain framework, providing a best practices framework for "service management". The nature of a service itself is defined in very broad terms as "a means of facilitating outcomes customers want to achieve without the ownership of specific costs" [OGC07]. In a telecommunications context, therefore, the scope can be construed to mean the lifecycle of a particular telecommunications service from inception to deletion.

The ITIL is in the public domain insofar as the published best practices are not proprietary. Crucially, however, ITIL is not prescriptive at a technology level. That is, it does not specify particular technologies, protocols, etc. to be used in the execution of its aims. Rather, it specifies a high-level framework but stops short of further specificity.

In its present form (version 3), ITIL is divided into several *Core Guidance Topics* comprising the overall *Service Lifecycle*. Their arrangement is shown in Figure 3-1 below. Particularly important for our purposes are the *Service Operation* and *Service Design* aspects of this lifecycle. Within these particular Core Guidance Topics resides process definitions that set the context for the protocols and technologies discussed later in this chapter.

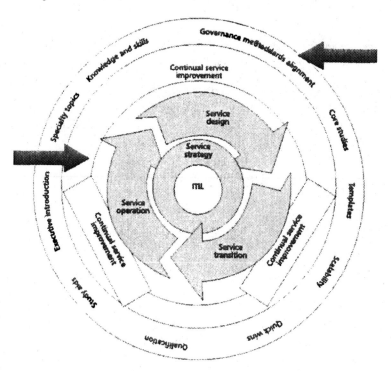

Figure 3- 1: The ITIL "Service Lifecycle"

3.3.1 Service Operation – the Relationship of Incidents, Problems, and Known Errors

Consider a hypothetical telecommunications company, complete with customers and their service expectations. The *Service Operation* aspect of ITIL proposes a series of processes and concepts for managing the performance of such services within their customers' expectations.

The first concept is that of the *Incident*, which ITIL defines as any event that disrupts, or could disrupt, a service [OGC07]. Such events are quite common in the life of telecommunications companies, and hence an *Incident Management* process is required to ensure the goal of "restoring normal service operation as quickly as possible and minimizing the adverse impact on business operations" [OGC07]. It is important to note here that the Incident itself is the occurrence of a service discontinuity or disruption specific to one or more end users. It is not necessary to understand the root cause of the Incident in order to fully describe it. Indeed, ITIL conveniently delineates the matter through the definition of a second concept: the *Problem*.

An ITIL Problem is defined to be "the unknown cause of one of more incidents" [OGC07]. There are two particular prerequisites implicit in this definition. First, it is impossible to have a Problem without also having one or more associated Incidents. This prerequisite gives the Problem its operational context; it is the undesirable condition that gives rise to service disruptions. Second, it is important to note the unknown nature of its root cause. This may seem a little disingenuous, but once a Problem's root cause is known, it is no longer a Problem. Rather, at this point, following successful diagnosis of its root cause, the Problem becomes what is defined as a *Known Error*.

Known Errors are essentially the determined root causes of previously established Problems. They may also, optionally, have a work-around associated with them (as may Problems themselves) to alleviate their disruptive effects in the short term until a more permanent solution can be devised and implemented. This permanent solution usually connotes a *Change*, which is a concept we will come to in the next section.

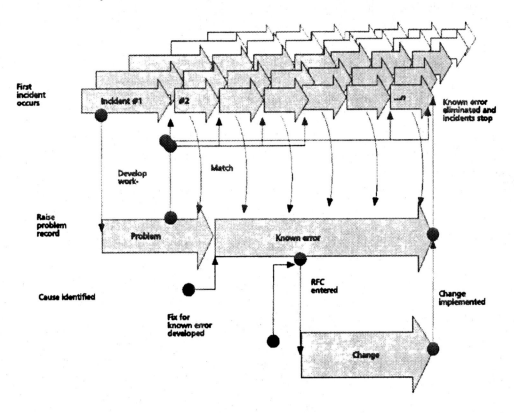

Figure3-2: The Relationship of Incidents, Problems, Known Errors and Changes

Figure 3-2 shows a simple example of the interaction between these four major ITIL concepts. This occurs in four discrete steps:

1. Incidents pertaining to a common root cause start to occur, and after some time operational staff notice the commonality and respond by raising a Problem record. From that point onwards, the

Incident Management process continues to recognize and record new incoming Incidents and match them to the Problem record.

2. The *Problem Management* process determines a work-around, which can be passed back to the customers reporting correlated Incidents. At some point later, the root cause is authoritatively diagnosed, a Known Error record is created, and the Problem record is associated with it.

3. Farther along in the process, a more permanent solution is identified to resolve the Problem in question. This involves the introduction of a Change via a mechanism that will be discussed in some detail shortly.

4. The introduction of this Change redresses the Known Error, and, therefore, the preceding Problem. Any preceding Incidents are similarly resolved at this time, and no further Incidents with that root cause recur in the future.

3.3.2 Service Transition – Change Management

Whereas Service Operation pertains to services that are in their steady state, *Service Transition* focuses on the procedures ITIL specifies for modifications to telecommunications services. This aspect of ITIL describes the procedures that should be employed when managing planned activities. The particular sub-aspect *Change Management* has a very specific bearing on the management of wireless networks. ITIL's definition of a *Service Change* is quite broad: "The addition, modification, or removal of an authorized, planned, or supported service or service component and its associated documentation" [OGC07-2].

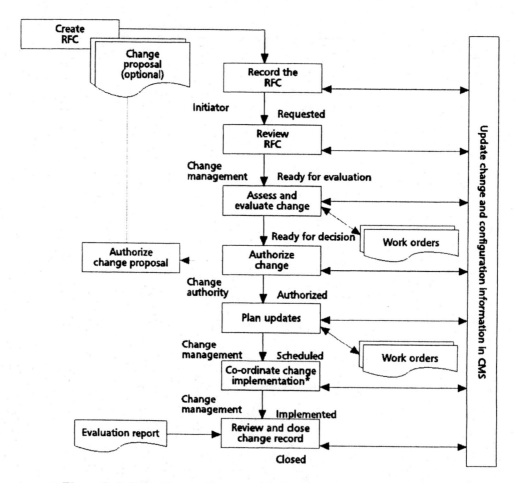

Figure 3- 3: ITIL Change Management Process for "Normal" Changes

This definition can be reasonably construed to encompass the modification of any element, however insignificant, making up the provision of a service to a customer. The ambiguity is arguably deliberate, however, as it allows the organization in question to determine the scope to some extent.

Change Management under ITIL is then the process through which the implementation of such Changes is controlled. A key consideration is the continuity or stability of the business, as change activity itself has been shown to be a significant driver of Incidents. Changes also need to be executed in an expedient manner, originating as they often do out of a need to resolve Known Errors, and out of a need to build new capabilities into existing infrastructure. The Change Management process therefore adds value by facilitating such modifications, while managing and controlling the inherent risks. Understanding that change activity varies greatly in terms of urgency and risk, the most recent version of ITIL proposes three variants of the Change Management process:

- The *Normal Change Model* is the baseline, depicted in Figure 3-3, from which the other models derive.
- The *Standard Change Model* is invoked in the case of frequently repeated Changes. In such situations, the risk is generally well understood and very low, and hence the *assessment* and *authorization* stages present in the Normal Change Model are deemed superfluous and are omitted as such.
- The *Emergency Change Model* is invoked in the case of urgent Change activity that needs to be applied immediately in order to mitigate an actual or imminent major service disruption. Incident or Problem activity is a frequent antecedent, in the manner described in Figure 3-2.

The Normal Change Model as described in Figure 3-3 commences with the creation of a *Request for Change* (RFC). This document will generally be required to contain a description of the change process, along with various other metadata, including but not limited to an estimate of any possible disruptions, the reasons motivating the Change, and contingencies in the event of failure [OGC07-2].

The assessment stage refers to an independent verification of the risks associated with the requested change. This is sometimes lumped in with the authorization stage, whose primary purpose is to weigh the relative risks and benefits of the request and then recommend either its approval or its rejection. Duly *authorized* or *approved* changes are then *scheduled* for implementation, and then *executed* by personnel. Finally, they are formally *closed*. This workflow is commonly managed by the company in question by a dedicated software system.

3.4   The Enhanced Telecom Operations Map

The *Enhanced Telecom Operations Map* (eTOM) model [TMF07] defines a set of business processes for telecommunications. This includes all aspects of running a large telecommunications company, but also, and importantly, processes relating to network management, monitoring resource performance, analyzing service quality and resolving customer-impacting incidents. eTOM follows a hierarchical decomposition of processes into levels. The first level (L0) defines three process areas and horizontal groupings based on the level of interaction with the customer. The three process groupings, shown in Figure 3-4, are:

- *Strategy, Infrastructure, and Product*, which includes processes that develop strategies and commitments to them within the enterprise.
- *Operations*, which includes all operational processes that support the customer and network management.
- *Enterprise Management*, which includes basic business processes required to run and manage large businesses: Financial Management, Human Resources, etc.

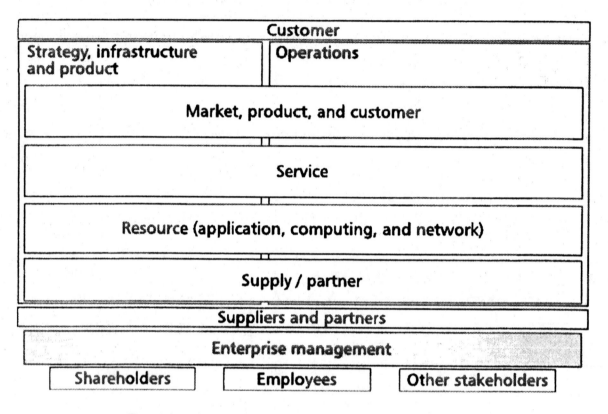

Figure 3- 4: Level Zero (L0) Decomposition of eTOM Processes

Operations processes are at the core of network management. These processes are technology and product agnostic and can readily be applied to the domain of wireless communication. The first level of these processes (L0) is divided vertically, as shown in Figure 3-5, to define the day-to-day operations of maintaining a customer service, service configuration, fault handling, and billing management:

- *Fulfillment* addresses customer order handling, service configuration, activation, and resource provisioning.
- *Assurance* addresses problem handling, quality of service, fault troubleshooting, and performance monitoring.
- *Billing* addresses bill creation, customer data records, collection, and rating.
- *Operations Support & Readiness* encompasses the additional processes required to facilitate the above three areas of the business.

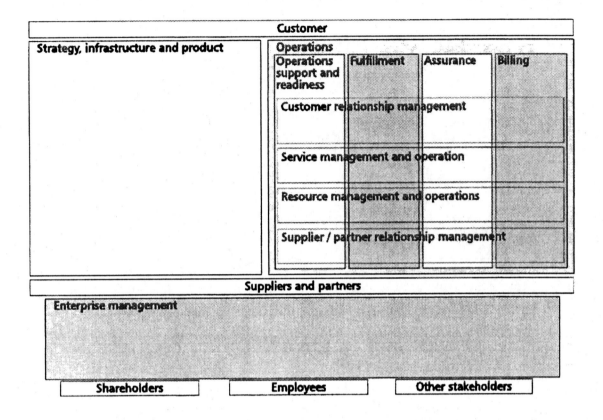

Figure 3- 5: Level One (L1) Decomposition of eTOM Processes Within *Operation*

The most relevant section to network management is Assurance. Each section is also divided horizontally (as with L0, shown in Figure 3-4) to define the level of interaction with the customer. These groupings specific to Assurance, shown in Figure 3-6, are:

- "Customer Relationship Management"–creating trouble tickets, track and manage customer SLAs.
- "Service Management and Operations"–diagnose service problems, monitor service quality.
- "Resource Management and Operations"–monitor resource performance, control quality of service, and
- "Supplier/Partner Relationships Management"–delineation of fault responsibility on network boundaries.

The next level of detail (L2) identifies the first specific processes that can be addressed by Operational and Business Support Systems (OSS/BSS). These processes can be used to make sure that current OSS/BSS have adequate coverage when introducing a new Product/Service and implementing the necessary Network Management to maintain that Product/Service. The eTOM vocabulary states that a Product (e.g., Mobile TV) is made up of one or more Services (e.g., Video Streaming, Customer Subscriptions) and enabled by many Resources (e.g., IT infrastructure, Streaming Servers, 3G Network).

Assurance		
Customer relations management	**Problem handling** Manage problems reported by customers associated with purchased product offerings. Receive reports from customers and provide repair status to the customer.	
	Customer QoS/SLA management Monitoring and reporting of delivered vs. contactual Quality of Service (QoS), as defined in the customer contracts or the catalogue of product offerings.	
Service management and operations	**Service problem management** Management of problems associated with a service. Respond immediately to reported service problems to minimize effects on customers.	
	Service quality management Monitoring, analysing and reporting performance of specific services. Responsible for restoring service performance to a level specified in the SLA.	
Resource management and operations	**Resource trouble management** Management of troubles associated with specific resources. Isolate the root cause and act to resolve the resource trouble.	
	Resource performance management Monitoring, analysing and reporting performance of specific resources from basic information received from Resource Data Collection processes.	
Supplier/partner management	**Supplier / Partner reporting management** Monitor and report problem engagements with Supplier / Partners to ensure interactions are in accordance with commercial arrangements.	
	Supplier / Partner performance management Monitor and report on service provider initiated performance engagements; higher level technical support (eg. product manufacturer), products/services purchased form an external supplier (eg. interconnect service).	

Figure 3-6: Level Two (L2) Decomposition of eTOM Processes Within *Assurance*

At level 3 (L3), Assurance is broken down into the sub-processes outlined in Figure 3-7. These sub-processes are linked within their corresponding (L2) processes by flow diagrams. There is also an inter-relation between (L2) processes in a way that defines a necessary course of action that traverses Customer, Service and Network boundaries; for example, to address a Customer Problem with a Service we "Create Customer Trouble Report", "Survey and Analyze Service Problem", "Localize Resource Trouble", and so on.

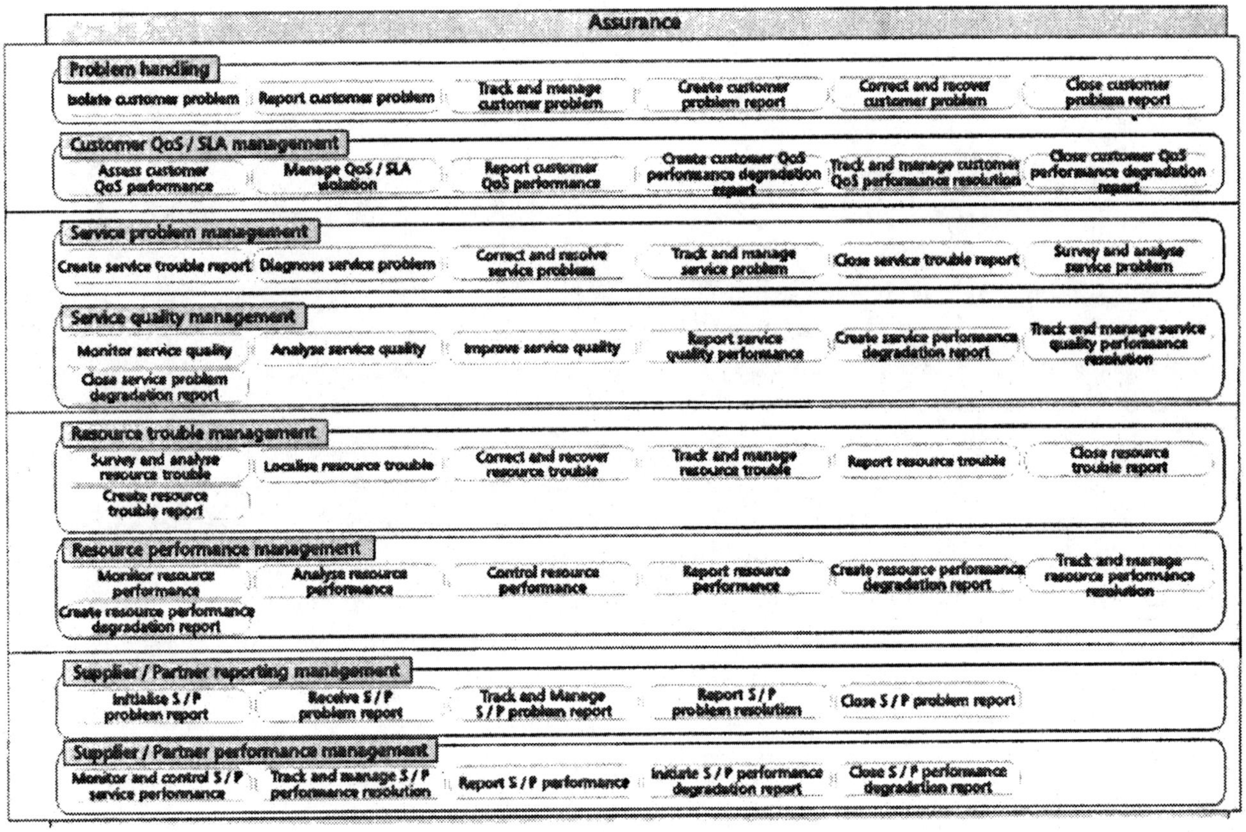

Figure 3-7: Level Three (L3) Decomposition of eTOM Processes Within *Assurance*

3.5   The Simple Network Management Protocol (SNMP)

Having established the operational context through discussion of ITIL and eTOM it is now meaningful to discuss the protocols and technologies that translate best practice into reality. One of the most significant protocols is the *Simple Network Management Protocol* (SNMP), which was originally defined through the Internet Engineering Task Force (IETF) beginning in 1988, and which has been in operation ever since. It is, as the name suggests, a relatively simple protocol, insofar as it had four defined operations in its initial formulation [RFC1157].

It is also worth noting at the outset that the protocol forms a part of the Internet Protocol (IP) suite, given that its messages are encapsulated within User Datagram Protocol (UDP) datagrams that are encapsulated within IP packets. This fact limits the applicability of SNMP to IP elements, although with the expansion of IP reachability within UTRANs, etc., the effective scope of the protocol is expanding significantly.

Central to the architecture inherent to SNMP is the notion of two distinct elements:

- The *SNMP Agents*, of which there are several, and possibly quite a large number. The Agents reside on the devices or nodes under management, and generally do three things. They service requests for information (in the form of "Objects") originating from the SNMP Manager pertaining to the node. They also respond to requests from the Manager to set a particular condition. In addition, they periodically send unsolicited messages to the Manager in reaction to a particular predefined condition or threshold being reached.
- The *SNMP Manager*, of which there is generally one single instance that receives information pertaining to the state

Version 1 of the SNMP protocol [RFC1157] specifies four basic operations, which are retained in present implementations. The first three pertain to the retrieval and setting of information on a network element under management. Note that the information in question is arranged in an object hierarchy, which will be described in some detail shortly.

- *get*, which is an instruction initiated by the Manager towards a specific network element (identified by its IP address). The agent is commanded to return the value of a specific Object retained on the Agent. This instruction, like the one following, is commonly issued by the Manager in order to determine a performance statistic or something similar.
- *get-next*, which is a partially stateful version of the previous instruction, aimed at traversing the tree-structure of the information retained on the Agent. When iterated, this allows the Manager to obtain all the values in the Agent's information base (which will be described shortly).
- *set*, which is an instruction initiated by the Manager towards a specific network element, which specifies that a particular Object on the Agent be set to a particular value. This instruction is commonly issued in order for the Manager to configure the node under management.

The remaining operation originates from the Agent and is the only one of the four that is initiated by the Agent:

- *trap*, which is a message sent from the Agent to the Manager advising that a particular pre-defined condition has been reached. These tend to be element-specific; however, there are several standardized traps, which will be discussed later.

More recent versions of SNMP use much the same model but offer some improvements. For SNMP v2 and v3, the basic differences with SNMP v1 are:

- SNMP v1 offers a simple request/response protocol and the operations listed above. The only security measure implemented is community strings, which have been shown not to be particularly effective.
- SNMP v2 offers additional protocol operations and multiple requested values (GetBulk and Inform), but still lacks strong security attributes. The definition of SNMP v2 can be found in RFC 1901.
- SNMP v3 adds security and remote administrative capabilities.

The SNMP v3 header has both security fields and non-security related fields. SNMP v3 provides a more secure level of authentication and privacy, and uses Data Encryption Standard (DES) encryption to encrypt the data packets.

The definition of SNMP v3 can be found in:

RFC 3411, Architecture for Describing SNMP Management Frameworks
RFC 3412, Message Processing and Dispatching
RFC 3413, Various SNMP Applications
RFC 3414, User-based Security Model (USM), providing for both Authenticated and Private (encrypted) SNMP messages
RFC 3415, View-based Access Control Model (VACM), providing the ability to limit access to different *Management Information Base* (MIB) objects on a per-user basis

The differences among the various SNMP versions are fully described in RFC 3584

3.5.1 The Structure of Management Information and the Management Information Base

Having discussed the basic interactions between the Agent and Manager elements, we now turn our attention to the structure of the information *stored within these elements* [Zel99].

First there is the Structure of Management Information (SMI) [RFC1155], which defines how information, arranged as "Objects" is to be named and defined. Each piece of information, or Object, has a specific place in a hierarchical tree structure. This particular identity is called the OBJECT IDENTIFIER; a series of integers which authoritatively names the Object. Consider the following example:

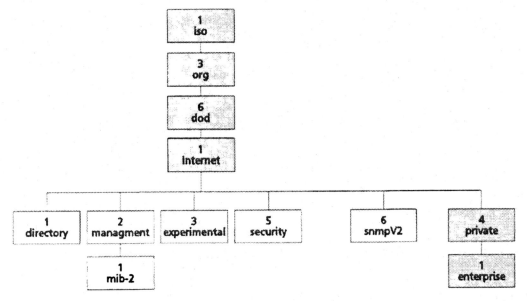

Figure 3- 8: Example Showing the Naming Convention for the MIB

Under the SMI scheme, the OBJECT IDENTIFIER for the *enterprise* Object in Figure 3-8 can be represented as:

{iso (1) org (3) dod (6) internet (1) private (4) enterprise (1)}

or by shorthand as

1.3.6.1.4.1

Now that we know how to address or name a particular Object in the hierarchy, we turn our attention to the types of data that can appear in such a hierarchy. RFC 1155 refers to these data generically as "Managed Objects" and defines a syntactical representation of these called the OBJECT-TYPE. Any such definition comprises five fields:

- *OBJECT*, which is a textual name for the managed object, along with its OBJECT IDENTIFIER denoting its place in the hierarchy.
- *SYNTAX*, which is an integer denoting the *type* of data wrapped by the managed object. Generally, only primitive types available in Abstract Syntax Notation One (ASN.1) are valid.
- *Definition*, which is a textual description, usually in ASCII, of the OBJECT TYPE's function.
- *Access*, which is an enumerated type specifying the read/write status of the object.
- *Status*, which is also an enumerated type specifying whether implementation of this object is required in order to claim full compliance of a MIB group.

Therefore, an OBJECT TYPE, named sampleObject, residing as the first leaf under the enterprise Object drawn in Figure 3-8, might be defined in the following ASN.1 notation:

```
sampleObject OBJECT-TYPE
        SYNTAX DisplayString (SIZE (0.255))
        ACCESS read-only
        STATUS mandatory
        DESCRIPTION
                "This is a sample OBJECT TYPE. Nothing more."
        ::= { enterprise 1 }
```

Readers are encouraged to familiarize themselves with SNMP v2 and v3. These are described in RFCs 1441-1452 and 3411-3418, respectively.

3.6 Security Requirements

The objective of communication security is the preservation of three principles: Confidentiality, Integrity and Availability. [ISO27002]

- Confidentiality: the communication data are only disclosed to authorized subjects.
- Integrity: the data in the communication retain their veracity and are not able to be modified by unauthorized subjects.
- Availability: authorized subjects are granted timely access and sufficient bandwidth to access the data.

The security mechanism that is needed is determined from the threat type and the affected principle. No one mechanism is able to address all such principles, so combinations of mechanisms are common. Wireless infrastructure has many common threats [NIST SP-800-48], the major categories of which are listed in Table 1.

Threat Category	Description	Principle affected
Denial of service	Attacker prevents or prohibits the normal use or management of networks or network devices.	Availability
Eavesdropping	Attacker passively monitors network communications for data, including authentication credentials.	Confidentiality
Man-in-the-middle	Attacker actively intercepts the path of communications between two legitimate parties, thereby obtaining authentication credentials and data. Attacker can then masquerade as a legitimate party.	Confidentiality/ Integrity
Masquerading	Attacker impersonates an authorized user and gains certain unauthorized privileges.	Integrity
Message modification	Attacker alters a legitimate message by deleting, adding to, changing, or reordering it.	Integrity
Message replay	Attacker passively monitors transmissions and retransmits messages, acting as if the attacker were a legitimate user.	Integrity
Traffic Analysis	Attacker passively monitors transmissions to identify communication patterns and participants.	Confidentiality

Table 1: Common Threats to Wireless Infrastructure

Other concepts that are important to know include:

- Identification: This occurs when a user (or client) presents himself to a system to claim his identity.
- Authentication: The mechanism by which the system verifies the identity of a user (or client).
- Authorization: The mechanism by which the system verifies that the user (or client) is permitted to execute a particular action.
- Accountability: The ability of a system to determine the actions and behavior of a user (or client).
- Privacy: The level of confidentiality and privacy protection given to a user by a system.
- Non-Repudiation: The user (or client) cannot deny that he is the individual who made a particular transaction.

3.6.1 Network Access Control

Physical access control poses a special problem, particularly given the near-ubiquity of the air interface of some wireless systems. Given that this ubiquity is often one of the key design objectives, *logical* access control is the mechanism used to achieve the required security.

Network access control is based on the procedure described before: to authenticate, authorize and account for a user or client. There are many approaches to managing these issues, and it is with this understanding that we describe some of the enabling technologies.

3.6.1.1 RADIUS

RADIUS (*Remote Access Dial In User Server*) is a standard protocol adopted by the IETF in RFC 2865 and RFC 2866, for carrying *authentication, authorization, accountability,* and *configuration* information between a *Network Access Server* and a centralized *Authentication Server* [RFC2865]. RADIUS's basic operation is shown in Figure 3-9.

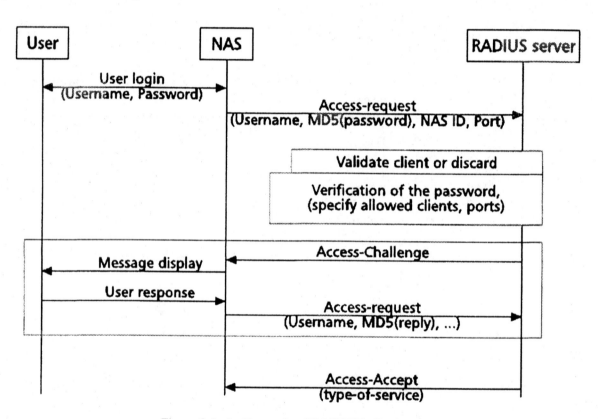

Figure 3-9: An Example of RADIUS's Operation

In this example, the end user presents their credentials to the Network Access Server (NAS). The NAS generates an Access-Request message to the RADIUS server. Once the RADIUS server receives the request it generates a challenge, and sends it back to the user. The user then performs a mathematical operation using the challenge (e.g. a hash) and sends it back to the RADIUS server. The RADIUS server then checks the credentials and generates either an Access-Accept or an Access-Reject message to the NAS, which itself confirms or denies the connection to the end user. In wireless networks, the NAS may be implemented within a specific access point such as the PDSN, GGSN, etc.

Is important to note that this is just the basic operation of the RADIUS protocol; all detailed options are described in RFC 2865. RADIUS packets are carried over UDP, and have many attributes (e.g., username, password, etc); the complete list is available in RFC 2865. There is also a capability for some attributes to be vendor-defined.

3.6.1.2 Diameter

Diameter is an Authentication, Authorization and Accounting protocol based on RADIUS for applications such as network access or IP mobility. Diameter is defined in RFC 3588 and is seen as a successor to RADIUS.

The primary ways in which Diameter improves upon RADIUS are [RFC3588]:

- Failover: RADIUS does not define failover mechanisms. Diameter supports application-layer acknowledgment and defines failover algorithms.
- Confidentiality: RADIUS defines the application-layer authentication and integrity scheme but not confidentiality (just optional support for Internet Protocol Security–IPSEC). By contrast, IPSEC support is mandatory for Diameter and TLS is optional.
- Reliable transport: To provide well-defined transport behavior, Diameter runs over the Transmission Control Protocol (TCP) or STCP protocols.
- Server-initiated messages: Support for server-initiated messages is optional in RADIUS, making it difficult to implement unsolicited disconnection or re-authentication/re-authorization across heterogeneous deployments. This support is mandatory in Diameter.
- Auditability: RADIUS does not define data-object security mechanisms and because of that, untrusted proxies may modify attributes or packet headers without being detected. The implementation of security objects is not mandatory in Diameter, but it is supported.
- Capability negotiation: RADIUS does not support error messages, capability negotiation, or mandatory/non-mandatory flags for attributes (AVPs). Diameter does provide this support.
- Peer discovery and configuration: Diameter enables dynamic discovery of peers using the Domain Name System (DNS). In addition, the derivation of dynamic session keys is enabled via transmission-level security.
- Roaming support: Diameter supports user roaming. By contrast, the IETF Roaming Operations (ROAMOPS) Working Group investigated whether RADIUS could be used to support roaming and concluded that it was ill-suited to handle the interdomain exchange of user and accounting information.

3.6.1.3 Extensible Access Protocol

In 1998, the Extensible Access Protocol (EAP) was defined by the IETF in RFC 2284. Subsequently, in June 2004 the technology was revised in RFC 3748. RFC 2284 focuses primarily on the message types and packet format. The security considerations and interactions with other protocols were addressed in the latter RFC [NIST SP800-97].

EAP may be used on dedicated links as well as switched circuits, and wired as well as wireless links. To date, EAP has been implemented with hosts and routers that connect via switched circuits or dial-up lines using Point to Point Protocol (PPP) [RFC1661]. It has also been implemented with switches and access points using IEEE 802. EAP encapsulation on IEEE 802 wired media is described in IEEE-802.1X, and encapsulation on IEEE wireless LANs in IEEE-802.11i [RFC3748].

In EAP, the party that wants to be authenticated is called the *supplicant* and the party that demands proof of authentication is called the *authenticator*. Four types of messages are defined in EAP: *request, response, success,* and *failure*. The authenticator sends a request message to the supplicant asking for a response message to authenticate. If the authentication is successful, a success message is sent to the supplicant; if not, a failure message is sent.

EAP supports many authentication methods, including passwords, digital certificates, tokens, etc. It is possible to combine these methods and use asymmetric methods for mutual authentication. The pass-through feature enables authenticators to forward the messages directly to the authentication server, offering the ability to vendors to generate new specific methods. These EAP methods perform the authentication transactions and generate the key material used to protect subsequent communications. RFC 3748 defines three authentication methods:

- MD5-Exchange (mandatory),
- One-Time-Password (OTP) (optional), and
- Generic Token Card (GTC) (optional).

However, none of these methods provides key material, and so is not suitable for WLANs. This is because WLANs use Transport Layer Security (TLS) -based EAP. RFC 3748 recommends that key establishment and generation should be based on mature and well-established techniques. TLS has thus emerged as the dominant protocol for EAP methods that support WLANs. Common TLS-based methods are:

- EAP-TLS,
- EAP Tunneled TLS (EAP-TTLS),
- Protected EAP (PEAP), and
- EAP Flexible Authentication via Secure Tunneling (EAP-FAST).

A description of each of these methods is beyond the scope of this chapter; the reader is referred to the RFC for details.

3.6.1.4 IEEE 802.1x

The broad success of Local Area Networks, both wired and wireless, and their growth in size has made requirements for security essential and the designation of a specific security standard within IEEE 802 desirable. In 2001, IEEE approved standard 802.1x, which specifies the port-based network access control for wired and wireless networks. IEEE 802.1x uses EAP as an auxiliary protocol to transmit the authentication data. When EAP is transmitted on a LAN (e.g., IEEE 802.3, IEEE 802.5, IEEE 802.11, etc.), it is encapsulated by the Extensible Authentication protocol (EAPoL).

The following definitions are important in order to understand 802.1x:

- The supplicant is the client that wants to authenticate to the network.
- The authenticator is the Wireless Access Point (802.11) or Ethernet Switch (802.3) that facilities authentication for the supplicant.
- The authentication server is the element that determines whether a supplicant is authenticated or not; it is generally a RADIUS server.

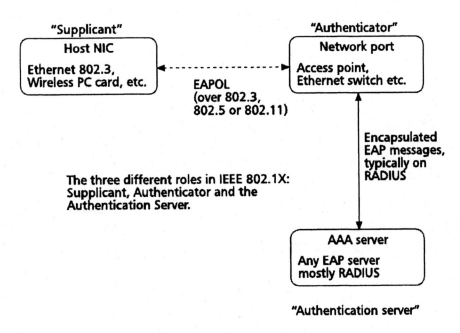

Figure 3-10: An Example of 802.1x Operation

Figure 3-10 shows the authentication of a user on an IEEE 802.11 network. The process starts when a non-authenticated supplicant tries to connect to the authenticator (the access point). The access point will only allow the client to generate EAP traffic (EAPoL) until it has been authenticated. The client sends the EAP-Start message, and the access point sends the EAP-Request Identity message to the client. The client answer is then sent to the RADIUS server to be verified. If verification is successful, the authenticator will then allow the client to generate any kind of traffic.

3.6.2 Wireless LAN Security

To be able to communicate data on a wireless network, a wireless station (STA) must establish an association with the access point (AP). Only after that association is established is an exchange of data allowed. In the infrastructure mode, the association process proceeds as follows [Arb01]:

- Unauthenticated and unassociated.
- Authenticated and unassociated.
- Authenticated and associated.

The original standard IEEE 802.11 defines two mechanisms to authenticate the access to the wireless LAN: open-system authentication and shared-key authentication, neither of them secure[1]. The shared-key authentication scheme, based on a unilateral challenge-response mechanism, is usually referred to as Wired Equivalent Privacy (WEP), which is described below. Open-system authentication is actually a *null* authentication mechanism, as it does not provide true identity verification. The client is authenticated to the AP having provided the following information:

---

[1] WEP encrypts the response by XOR'ing the challenge with a pseudo-random key stream generated using a WEP key. The attacker can XOR the challenge and the response to expose the key stream, which can subsequently be used to authenticate [NIST SP800-97].

- Service Set Identifier (SSID): The SSID is the name assigned to the wireless LAN used by the client to identify the network. The SSID is generally transmitted as plain text, although a security control can generally be set to not broadcast it.
- Media Access Control (MAC) address: Many wireless LAN implementations allow the administrator to specify the MACs allowed to access the network using access lists. This too is an imperfect precaution, as the MAC is also sent as plain text and is thus susceptible to spoofing.

Shared key authentication is based on a common secret shared between the access point and the client, a cryptographic key known as a WEP key. It uses a simple challenge-response scheme whereby the STA initiates an Authentication Request with the AP, and the AP generates a random 128-bit challenge. Once the client receives the management frame from the responder, it copies the contents of the challenge text into a new management frame body. Using the WEP key and the Initialization Vector (IV), the STA encrypts this new management frame and sends it back to the AP for verification. If the verification at the AP is successful, the AP and STA switch roles and the process begins again to ensure mutual authentication. The procedure is shown in the Figure 3-11 [Arb2001]. An additional limitation of shared-key authentication is that it only authenticates the client, not the actual user.

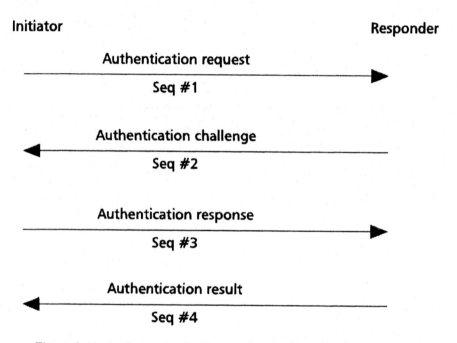

Figure 3-11: An Example of a Shared Key Authentication Operation

3.6.2.1 Wired Equivalent Privacy

The *Wired Equivalent Privacy* (WEP) protocol was designed to provide confidentiality for wireless network traffic. It is based on the RC4[2] cipher algorithm. The standard defines 40-bit and 104-bit keys; however, many vendors have implementations with longer keys (128-bit, 256-bit, etc). WEP also uses a 24-bit Initialization Vector (IV) as a seed value to initialize the cryptographic system. For example, a 40-bit WEP key and an IV are used to generate a 64-bit key for the system. RC4 is a stream cipher algorithm that generates a pseudo-random sequence, which is unified with the message using an XOR operation. Of course, a different sequence is needed for each message; that is where the IV becomes

---

[2] RC4 does not meet FIPS requirements for cryptographic algorithms. Accordingly, some government agencies and others requiring FIPS-validated solutions cannot use WLAN security solutions based on RC4, including both TKIP and WEP. I

significant. The IV's short length was one of the main weaknesses of the protocol, as evidenced by the many open-source tools available today which are capable of attacking it.

For integrity checking, WEP uses a 32-bit Cyclic Redundancy Check (CRC-32). This is computed for each payload before encryption takes place. The CRC-32 is then encrypted and becomes the Integrity Check Value (ICV) on the frame to be transmitted. CRC-32 is vulnerable to bit-flipping attacks, allowing an attacker to know which bits change when message bits are modified. The bit-flipping flaw persists after a stream-cipher step such as RC4. For this reason, WEP protocol integrity control is vulnerable to bit-flipping attacks.

3.6.2.2 Temporal Key Integrity Protocol

The Temporal Key Integrity Protocol (TKIP) was designed to strengthen the WEP protocol without causing significant performance degradation. It mainly uses the same algorithms as WEP and because of that, no hardware replacement is often required for an equipment upgrade from WEP to TKIP. WiFi Protected Access (WPA) certification, from the WiFi Alliance, requires the implementation of TKIP. TKIP provides the following security features which improve on WEP [NIST SP800-97]:

- Confidentiality protection, using the RC4 algorithm,
- Integrity protection, using the Message Integrity Code (MIC) based on the Michaels algorithm,
- Replay prevention, using a frame sequencing technique, and
- Use of a new encryption key for each frame.

TKIP uses three keys during the encapsulation process, two of them for integrity control (each half duplex channel) and the other for encryption. Table 3-2 lists the primary characteristics of TKIP and the security principle achieved:

TKIP Characteristic	Security Principle
Two 64-bit message integrity keys are used for MIC. The MIC is computed over the user data, source and destination addresses, and priority bits.	Integrity
A monotonically increasing TKIP Sequence Counter (TSC) is assigned to each frame. This provides protection against replay attacks.	Integrity
A key-mixing process produces a new key for every frame. It uses the Temporal key (explained later) and the TSC to generate a dynamic key	Confidentiality
The original user frame, the MIC and the source address are encrypted using RC4 using the per-frame key.	Confidentiality

Table 3-2: TKIP Characteristics

3.6.2.3 Counter Mode with Cipher Block Chaining MAC Protocol

The Counter Mode with Cipher Block Chaining MAC Protocol (CCMP) was developed with many of the same objectives as TKIP; however, CCMP development was done with fewer hardware restrictions. It is based on a generic authenticated encryption block cipher mode of the Advanced Encryption Standard (AES) Counter with Cipher Block Chaining Message Authentication Code (CBC-MAC) (CCM)[3] mode. CCM is a mode of operation defined for any block cipher with a 128-bit block size. In few words, AES is to CCMP as RC4 is to TKIP [NIST SP800-97] [Leh06].

---

[3]   CCM is defined by RFC 3610, counter with CBC-MAC (CCM) (http://www.ietf.org/rfc/rfc3610.txt). AES is defined by FIPS PUB 197 (http://csrc.nist.gov/publications/fips/fips197/fips-197.pdf).

CCMP implementation is mandatory for WPA2 WiFi Alliance certification. In most cases, the cryptographic characteristics of CCMP require a hardware change for the upgrade from WEP.

CCMP generates an integrity control (MIC) using Cipher Block Chaining (CBC-MAC). The key used for encryption and integrity is the same, but with different initialization vectors. The integrity protection is done for the payload and header, while the encryption (confidentiality protection) is done only for the payload. This allows for easier detection of a wrong packet.

CCMP processing expands the original MAC Protocol Data Unit (MPDU) size by 16 octets, 8 octets for the CCMP Header field and 8 octets for the MIC field (see Figure 3-12). The CCMP Header field is constructed from the Packet Number (PN), ExtIV, and Key ID subfields. PN is a 48-bit PN represented as an array of 6 octets. PN5 is the most significant octet of the PN, and PN0 is the least significant. Note that CCMP does not use the WEP ICV [IEEE 802.11i-2004]. CCMP key space has size 2^{128} and uses a 48-bit PN to construct a cryptographic nonce, which is a number or bit string used only once in security engineering, to prevent replay attacks. The construction of the nonce allows the key to be used for both integrity and confidentiality without compromising either.

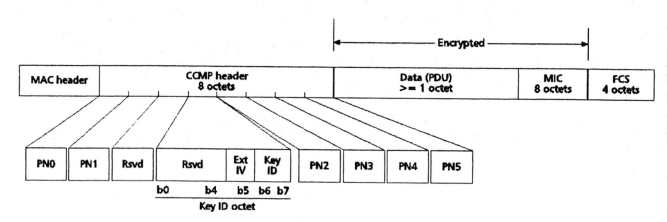

Figure 3-12: CCMP's Expanded MAC Protocol Data Unit (MPDU)

The encapsulation process is as follows [IEEE 802.11i-2004]:
- Increment the PN to obtain a new PN for each MPDU. The retransmitted MPDUs are not modified.
- Use the fields in the MPDU header to construct the additional authentication data (AAD) for CCM. The CCM algorithm provides integrity protection for the fields included in the AAD. MPDU header fields that may change when retransmitted are muted by being masked to 0 when calculating the AAD.
- Construct the CCM nonce block from the PN, A2, and the Priority field of the MPDU, where A2 is MPDU Address 2. The Priority field has a reserved value set to 0.
- Place the new PN and the key identifier into the 8-octet CCMP header.
- Use the temporal key, AAD, nonce, and MPDU data to form the cipher text and MIC. This step is known as CCM originator processing.

3.6.3 Robust Security Networks (IEEE 802.11i/WPA2)

Amendment IEEE 802.11i to 802.11 introduces the concept of Robust Security Networks (RSNs) and Robust Security Network Associations (RSNAs). These new concepts imply enhanced security features beyond the simple shared-key challenge-response authentication just described [NIST SP800-97]. In RSNAs, IEEE 802.1x provides authentication and controlled port services, while key management is done using IEEE 802.1x and IEEE 802.11 working together. All stations (STAs) in an RSNA have a corresponding IEEE 802.1X entity that handles these services [IEEE 802.11i-2004].

In addition to WEP and IEE 802.11 authentication, an RSNA defines [IEEE 802.11i-2004]:

- Enhanced authentication mechanisms for STAs,
- Key management algorithms,
- Cryptographic key establishment, and
- An enhanced data-encapsulation mechanism, called CTR (counter mode) with CCMP and, optionally, TKIP.

IEEE 802.11i offers two general classes of security capabilities for IEEE 802.11 WLANs. The first class, pre-RSN security, includes the security mechanisms explained before: open-system or shared-key authentication for validating the identity of a wireless station, and WEP for the confidentiality protection of traffic. The second class of security capabilities includes a number of security mechanisms to create RSNs [NIST SP800-97]. Figure 3-13 illustrates the security mechanisms for RSN and pre-RSN networks.

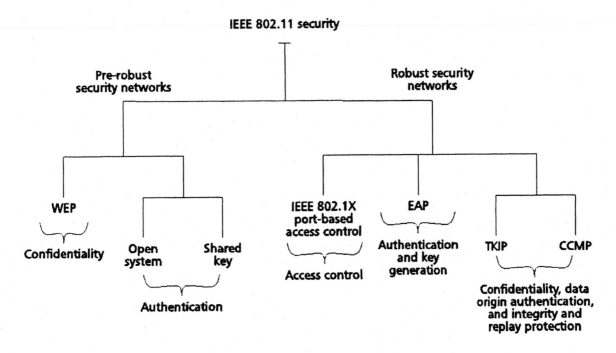

Figure 3-13: Hierarchy of 802.11 Security Mechanisms

3.6.3.1 Key Hierarchies

Among the most important cryptographic characteristics are the keys used during the process. Pre-RSNs use manual WEP keys; only one key is used and no key management exists. For RSNs, many keys are needed for integrity and confidentiality protection. IEEE 802.11i defines two key hierarchies for RSNs: the Pairwise Key Hierarchy (PKH) for unicast traffic and the Group Key Hierarchy (GKH) for multicast/broadcast traffic [NIST SP800-97]. In PKH, the process begins with one of the two root keys. From those keys, all key material is produced. These pairwise keys are used by both TKIP and CCMP. The Pairwise Master Key (PMK) is taken directly from the root key. This key has at least 256 bits, depending on the generation method (Phase Shift Keying–PSK or Adaptive Antenna Array–AAA).

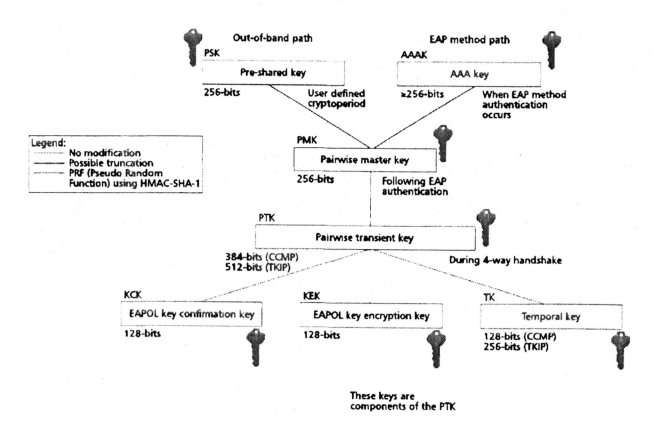

Figure 3- 14: Pairwise Key Hierarchy

As illustrated in Figure 3-14, in PKH, after the Pairwise Transient Key (PTK) is generated from the PMK using a pseudo-random function and keyed-Hash Message Authentication Code (HMAC-SHA1), the PTK is partitioned into the EAPoL Key Communication Key (KCK), the EAPoL Key Encryption Key (KEK), and the Temporal Key (TK). These keys are used to protect unicast communication between the authenticator's and supplicant's respective clients. PTKs are used between a single supplicant and a single authenticator [IEEE802.11i 2004].

The other key hierarchy is the Group Key Hierarchy. As shown in Figure 3-15, it consists of a single key: the Group Temporary Key (GTK). The GTK is generated by the authenticator and transmitted to all the stations.

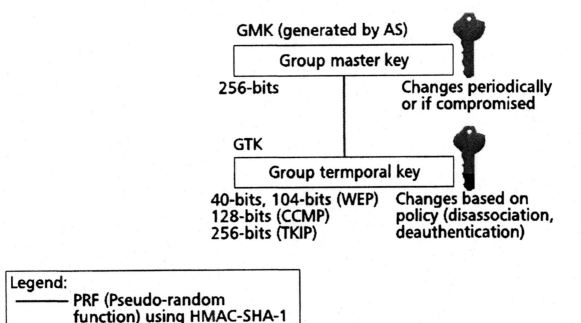

Figure 3-15: Group Key Hierarchy

3.6.4 3GSM Security

3GSM security principles are similar to other networks, such as WLANs. However, user expectations for fast communications and ease of use, as well as the propensity for handsets to be easily lost or stolen, present a number of unique challenges. Under the first-generation AMPS system, only the device's serial number was used to validate the handset. In short order, many devices were developed to compromise the system—for example, by intercepting serial numbers over the air and thus cloning handsets.

The second-generation GSM system was developed with security in mind. The responsibility for user security is largely in the hands of the Home Environment (HE) operator. The HE operator can control the use of the system by the provision of Subscriber Identity Modules (SIMs), which contain user identities and authentication keys.

The design of third-generation security was based on the following principles [TS 33.120]:

- 3G security will build on the security of second-generation systems. Security elements within GSM (e.g., SIMs) and other second-generation systems that have proved to be needed and robust will be adopted for 3G security.
- 3G security will improve on the security of second-generation systems: 3G security will address and correct real and perceived weaknesses in second-generation systems.
- 3G security will offer new security features and will secure new services offered by 3G.

3.6.4.1 GSM legacy

One of the most important elements of GSM security is the SIM. This removable security card is terminal-independent and requires almost no user-intervention (aside from the input of a PIN at device-boot time). The SIM is managed by the HE and contains all the identification data and cryptographic keys needed for the subscriber to make or receive a call. The authentication process is shown in Figure 3-16.

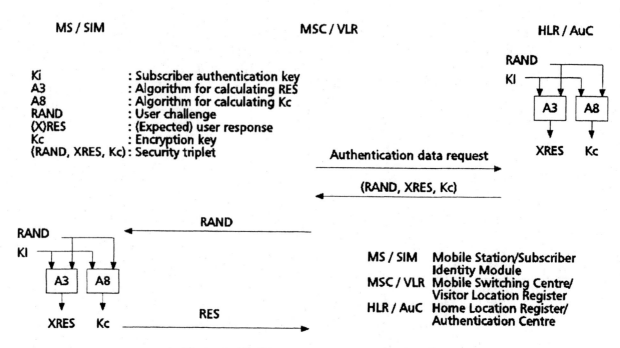

Figure 3- 16: GSM Authentication Mechanism

The Authentication Center (AuC) associated with the Home Location Register (HLR) contains the identification material for the user (i.e., the shared secret key). The Mobile Station (MS) attempts authentication to the network via the Mobile Switching Center/Visitor Location Register (MSC/VLR), which in turn passes the request to the HLR/AuC. It responds with a challenge (128-bit RAND), a key for encryption (Kc), and an expected value of the user expected response (XRES) to the MSC/VLR. The network sends the RAND to the MS, the MS transfers the parameter to the SIM, and the SIM generates the RES (32 bits) and Kc (64 bits) using the RAND and the shared secret key Ki. After that, the MS sends the response (RES) to the network and the network compares the RES to the XRES for verification. The functions that generate the RES and the Kc are called A3 and A8, respectively.

During call establishment, an encrypted mode of transmission is established where the MSC/VLR transports the current Kc to the base station and then instructs the MS to select the same Kc generated in the SIM during authentication. The encryption algorithm, called A5, is a stream cipher. Currently, three different A5 algorithms have been standardized, called A5/1, A5/2, and A5/3. The encryption process is done before modulation but after the interleaving process using a stream cipher (function A5) and the Kc. Note that this encryption is only for the air interface. Encryption of the channel is switched on or off by the base station, which also selects the algorithm in use.

The primary security issues affecting GSM are:

- The network is not authenticated by the MS, so impersonation attacks are possible using rogue network hardware.
- Only the air part of the communication is encrypted.
- Cipher keys and authentication values are transmitted unencrypted within and between networks (IMSI, RAND, SRES, and Kc).
- Protection against radio-channel hijacking and channel integrity rely on this encryption.
- Some of the algorithms have been proven weak through successful attacks.

3.6.4.2 3GSM Security Domains

The first part of the 3GSM security model is the concept of Security Domains. A Security Domain is a collection of elements that is under the control of one security authority, and performs a security function. In particular, the following domains are specified:

- Home Network (HN),
- Visited Network (VN),
- Access Network (AN), and
- Third-Party Application Service Providers (ASPs).

Note that the AN is not necessarily part of the VN, but could be a third-party entity such as an Internet café or a hotel network. In many cases, this network does not participate other than to provide carriage. The domains are illustrated in Figure 3-17. The two third-party application providers have relationships with the networks. However, the two could be the same provider. For example, a content downloader could have relationships with more than one carrier. Thus, a customer of one roaming on the other will be able to access content through either carrier. It is expected that there may be services from one carrier not available to another, for example a business locator in a particular city, which would be used by customers of both—e.g., lost tourists in a city.

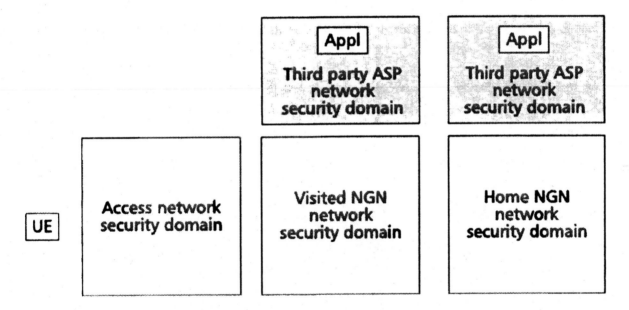

Figure 3- 17: NGN Security Domains

The AN is abstracted from the 3G network. This is the case in other standards, and it is useful to distinguish these networks for a number of reasons: it makes the security associations and relationships clearer, and it may be imposed by regulation and reselling. The standards assume that some sort of Security Gateway Function (SEGF) exists between the domains. The details of the security gateway are not specified within 3G, leaving it to the domains to determine their requirements.

3.6.4.3 Security Associations

Security domains can form Security Associations (SAs) between themselves. Two direct SAs are defined and are shown in Figure 3-18: between the User Equipment (UE) and the HN and between the HN and the VN.

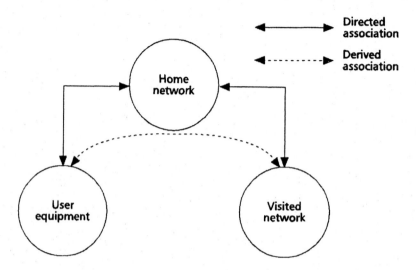

Figure 3-18: The Security Associations of UMTS/GSM

Under 3GSM, the UE is assumed to incorporate a SIM, which provides certain functions such as authentication and key agreement, and provides secure storage and processing. Note that the HN also has a security association with third-party ASPs. There is no direct relationship between the UE and the VNs. The other relationships form the basis of the model. Trust is established by the UE being authenticated to the network. The 3G Security Architecture is illustrated in Figure 3-19, which identifies the five SAs.

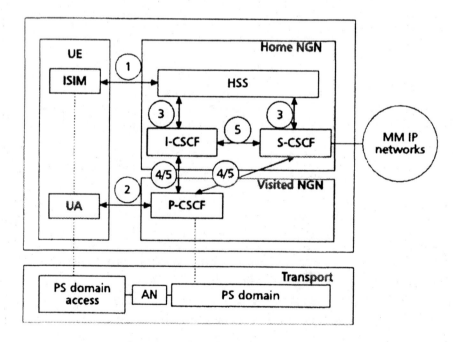

Figure 3-19: 3G Security Architecture

The standard explains this diagram using the following numerical keys.

1. This is the long term SA between the SIM and the HSS. It provides mutual authentication. Note that the authentication is usually carried out by the Serving Call Session Control Function (S-CSCF), although the HSS generates the challenges and expected responses.
2. This is the secure link between the UA and the Proxy-Call Session Control Function (P-CSCF). It supports data-origin authentication.
3. This is the security association for the internal elements of the network. This is more fully described in European Telecommunications Standards Institute (ETSI) TS 133 120.
4. This is the security association for Session Initiation Protocol (SIP) nodes. If the P-CSCF is in the Home Next Generation Network (NGN), then this relationship is covered by 5.
5. This is the internal security association for SIP nodes within the network.

Figures 3-18 and 3-19 show the explicit links for the home network and for the visited network. Note that in both of these cases the transport layer is assumed to be independent.

3.6.4.4 Trust Model

In the case where the UE does not have a relationship with the network, the home network provides the authentication vectors—that is, a set of challenges, the expected responses, the responses expected by the UE, and the cryptographic keys. Thus, the trust relationships use a delegated Trust Model. The user trusts the VN because the user trusts the HN, which has a trust relationship with the VN. This is the model used in mobile networks and roaming. This model is expected to be extended to wireline services.

Under GSM, the handset does not authenticate the network; it implicitly trusts the network equipment. This led to some security weaknesses that allowed the confidentiality of calls to be compromised by unauthorized third parties. The security weakness has been addressed by having the home and visited network provide responses to the handset. The handset compares these responses against the expected responses from the network. If the network fails this authentication, then the handset will not connect. An important aspect of this set up is that all signaling and control traffic is sent from the VN to the HN. This is used for billing and to validate the signaling from the UE.

3.6.4.5 Interconnect Security

The security domains are controlled by one security authority and are linked to each other by security gateways. The 3GSM standards do not prescribe what protocols or mechanisms should be used to connect these gateways or to protect the traffic. Consequently, carriers need to determine requirements for the connections between domains. Among these requirements should be:

- standards-based protocols, such as IP security
- strong authentication—probably certificate-based
- regular re-authentication
- use of standards-based cipher
- integrity mechanisms

The key lengths and algorithms should be reviewed regularly to ensure that they will continue to satisfy the security objectives. Initially, the security gateways should use IPSec with 2048-bit RSA public key modules for authentication, Internet Key Exchange (IKE) to provide key exchange and management, 128-bit Rijndael as the cipher between the gateways, and Message Digest algorithm 5 (MD5) providing integrity. The security gateway is expected to be a network gateway or edge router. Once the gateways are secured, the security domains trust each other. The risk of trusting other security domains is that fraudulent traffic may originate in the trusted domain. This may be due to a compromise within the domain or to malicious activity by operators of the domain. Thus, due diligence is required when establishing a trusted domain. There exist many commercial procedures to establish trust relationships. These include requirements that any trust domain should:

- comply with international risk-management standards such as ISO/IEC 17799
- have fraud-management processes
- have dispute-resolution procedures
- have procedures for carrying out investigations, including joint investigations with other carriers and ISPs

These, when used with other commercial procedures, will reduce exposure to fraud.

3.6.4.6 Authentication and Key Agreement

The fundamental protocol used within 3G for authentication and key exchange is the Universal Mobile Telecommunications System Authentication and Key Agreement (UMTS AKA). This is a zero-knowledge protocol. That is, no secret information is passed between two parties. It also provides mutual authentication of the UE and the network. There are three steps in the protocol.

- The UE requests registration and access.
- The network responds with a random challenge and a network authentication token.
- The UE validates the network's authentication token and responds with its own token.

After these exchanges, the network validates the response from the UE. If this response is validated, then the cryptographic keys are derived. If the network is a VN, then the VN passes the request to the HN, which responds by passing authentication vectors to the VN. The authentication vectors consist of the random challenge to be used, the expected response from the UE, the network authentication token, and the cryptographic keys. Note that at no point do either the UE or the HN pass on the shared secret.

3.6.4.6.1 The Keys and Other Parameters

The authentication key K is a fixed 128-bit key held within the SIM. It is the long-term secret key shared between the SIM and the AuC. See Table 3-3 for a summary of the size of this and the following parameters.

The random value RAND is generated by the serving network at the time of request.

The sequence number SQN is 48 bits and is used as a counter. This counter is kept by both the MS and the HN. It is a generated value, although there is no prescribed method to generate it. However, there are methods for maintaining synchronicity between the HN and the MS.

The Authentication Management Field (AMF) is a 16-bit parameter; it may vary from domain to domain. The only constraint is that the format and interpretation of the AMF must be the same for all implementations in a domain. Among the examples of the AMF are: support for multiple authentication algorithms and keys; changing sequence number verification parameters; and setting threshold values to restrict the lifetime of cipher and integrity keys.

All other values are derived from the four values just defined:

- The cipher key CK is the session key. It has 128 bits.
- IK is an integrity key. It has 128 bits.
- The message authentication code MAC is a 64-bit field. It is used in the authentication of the mobile device and the network. Note also that the parameter XMAC is used by the SIM to authenticate the network.
- MAC-S is an authentication token used in resynchronization and to construct AUTS.
- Authentication Token (AUTN) is a network authentication token.
- AUTS is a token used in resynchronization.
- AK is an anonymity key used to conceal the sequence number, as the latter may expose the identity and location of the user. The concealment of the sequence number is to protect against passive attacks only. If no concealment is needed then $f5 = 0$ (AK = 0).

Parameter Sizes										
K	RAND	SQN	AMF	CK	IK	MAC	MAC-S	AUTN	AUTS	AK
128	128	48	16	128	128	64	64	64	64	48

Table 3-3: Parameter Sizes

3.6.4.6.2 Cryptographic Primitives

Nine cryptographic primitives have been specified for use in 3G. These have been specified at two levels: the first with the parameters and lengths, and the second with the actual algorithms. This means that if a weakness is found with a particular algorithm, then another algorithm with the same parameters can be specified. However, if the length of authentication, for example, is no longer considered sufficient, then these parameters will need to be re-specified. These algorithms provide the functions used to:

- authenticate the UE and the network,
- derive cipher keys,
- derive anonymity keys, and
- resynchronize keys

The confidentiality and integrity ciphers are based on the block cipher Kasumi. Kasumi encrypts blocks of 64 bits and uses a 128-bit key. The functions for authentication and key agreement use a cryptographic primitive that is a block cipher with a 128-bit key and 128-bit block size. The standards do not specify which block cipher, only the parameters. The cryptographic primitives or functions used are:

- f1–Message authentication function used to compute MAC
- f1*–Message authentication function used to compute MAC-S
- f2–Message authentication function used to compute RES and XRES
- f3–Key generating function used to compute CK
- f4–Key generating function used to compute IK
- f5–Key generating function used to compute AK in session set-up
- f5*–Key generating function used to compute AK in re-synchronization procedures
- f8–Data confidentiality algorithm for the session (uses CK)
- f9–Data integrity algorithm for the session (uses IK)

More specifically:

$$
\begin{aligned}
\text{MAC} &= \text{f1(K, RAND, SQN, AMF)} \\
\text{XMAC} &= \text{f1(K, RAND, SQN, AMF)} \\
\text{RES} &= \text{f2(K, RAND)} \\
\text{XRES} &= \text{f2(K, RAND)} \\
\text{CK} &= \text{f3(K, RAND)} \\
\text{IK} &= \text{f4(K, RAND)} \\
\text{AK} &= \text{f5(K, RAND)}
\end{aligned}
$$

These are used in the protocols for authentication and key exchange. Thus the functional types of these are:

$$
\begin{aligned}
\text{f1}: &\ \{GF(2)\}^{320} \to \{GF(2)\}^{64} \\
\text{f3}: &\ \{GF(2)\}^{256} \to \{GF(2)\}^{128} \\
\text{f4}: &\ \{GF(2)\}^{256} \to \{GF(2)\}^{128} \\
\text{f5}: &\ \{GF(2)\}^{256} \to \{GF(2)\}^{48}
\end{aligned}
$$

The function f2 will be fixed within the domain; however, its output varies between 4 and 16 octets (32 to 128 bits). There may be other constraints, such as buffer size in the SIM or the packet sizes in the exchange.

3.6.4.6.3 Description of Protocols

There are two main protocols: the distribution of authentication data from the HE to the SN, and the authentication between the SN and MS. In the first protocol, the SN requests the authentication information from the HE. Essentially the IMSI (i.e., the permanent number) is sent from the SN to the HE. The HE responds with: {AV1, …,AVn}, where each AVi is defined:

AV: = RAND ‖ XRES ‖ CK ‖ IK ‖ AUTN, and
AUTN: = SQN ‖ AK ‖ AMF ‖ MAC.

Note that a fixed number of authentication vectors are transferred. This allows the SN to authenticate a number of times without reference to the HE. Now that the SN has the authentication vectors, the authentication protocol is straightforward.

- The SN sends the MS the request: RAND ∥ AUTN.
- The MS responds with RES.
- The SN compares RES with XRES. If they differ, authentication fails.
- The MS generates XMAC and compares it with MAC. If they differ, the MS rejects user authentication and sends an error code.
- If both the SN and the MS accept the authentication, then the session is established.

Note that the user authenticates the network. This mutual authentication overcomes a major shortcoming of the GSM protocol.

The Trust Model is essentially a delegated trust model. Trust is delegated by tokens. A network, N, will accept a registration attempt from UE that claims to originate either from N or from a home network (HN) with which N has a relationship. The UE will attempt to register with the network. N acts as a proxy for the registration messages. If the registration is successful, the HN will send a message to N indicating that the registration has been successful. This message is essentially confirmation that the HN will accept charges incurred by the UE. Similarly, when the UE registers, it will determine expected responses from N. If the responses received from the network correspond with the expected response, the UE will trust N. Note that in this way, both the UE and the network authenticate each other. Note also that in the roaming case, there is no direct security association between the UE and the HN.

3.6.4.6.4 Analysis of Security Protocols

For security analyses, a distinction is made between protocols and cryptographic primitives, such as ciphers. A protocol is a series of messages exchanged between two or more parties; the messages are processed and other messages are generated and decisions are made based on that processing. Cryptographic primitives are the functions used to generate and process messages and responses. In the protocol analysis, the cryptographic primitives are assumed to be secure.

AKA is the protocol that is used for authentication and key agreement. It is an extension of the authentication and key-agreement protocol used in GSM. In GSM the network is not authenticated, which allowed user confidentiality to be compromised by a rogue base station. This flaw has been addressed in 3G, and no exploit of AKA has been published. However, this does not mean that no exploit exists. For protocols, particularly key-exchange protocols, methods of evaluating and validating these protocols have been developed. Either a protocol logic, such as the Burrows-Abadi-Needham (BAN) logic, or model checking is used.

Køien has completed a formal mathematical analysis of the AKA protocol [Køi02]. In this model, the cryptographic primitives are assumed to be secure. The formal analysis found no flaw in the protocol and indicated that it was sound. A caveat indicates that the model may be incomplete. However, the protocol is simple and involves a small number of steps, so the model is accurately described in the formal language and therefore the probability that a flaw exists is small. Thus, we have confidence that the UMTS AKA protocol is sound. The security is, therefore, dependent on the security of the cryptographic primitives.

3.6.4.6.5 Analysis of Cryptographic Primitives

The cryptographic primitives can be grouped into two sets:

- primitives used for confidentiality and integrity of the traffic between the UE and the network, and
- primitives used for user authentication, key agreement, and generating nonces (random values).

Those in the first set are based on the Kasumi block cipher. Those in the second set use another kernel, or block cipher; the actual cipher is not specified, just its parameters. The standards use the Rijndael cipher in the examples, and the test data assume the use of this cipher.

Note that for authentication, key agreement and nonce messages are passed between the SIM and the HN, while the VN does not process these messages. Thus, the cryptographic primitives used by the HN and the SIM do not have to be shared with the VN, and therefore do not need to be standardized.

The 3G standards define the confidentiality and integrity functions (ciphers). These are based on the Kasumi block cipher. The other cryptographic primitives assume that a kernel is used; the kernel is assumed to be a block cipher with a 128-bit key and a 128-bit input/output. The example cipher used is Rijndael, the cipher chosen for the Advanced Encryption Standard (AES), and all the test vectors assume that Rijndael is used.

Unlike protocols, there is no way of being able to prove the soundness or security of a cipher or cryptographic algorithm [Col97]. The approach used by cipher designers has been to use strong structures and design criteria to avoid known weaknesses. These ciphers are then submitted to the cryptographic community for scrutiny.

The primitive Kasumi is a block cipher based on the Misty block cipher, designed by Mitsuru Matsui and published in 1995. The details of Kasumi were published in 1999. The most successful attack on Kasumi was published in 2005 by Biham, et al. [Bih05]. This attack requires $2^{54.6}$ chosen plaintext/ciphertext pairs (about 43 petabytes) and needs $2^{76.1}$ operations of Kasumi—the equivalent of encrypting the entire 43 petabytes two million times. This will recover one session key of Kasumi—the session key being updated regularly, at least daily. This attack is not feasible and does not appear to lead to attacks that are more efficient. Since this cipher is used to protect the RF interface, it is sufficiently secure for the medium term, say the next 10–15 years.

The 3G standards specify the other functions and primitives using a block cipher. The block cipher is not specified, only that it has a 128-bit key and a block size of 128 bits. The functions are built upon the cipher. The cipher used as an example in the standards is Rijndael, the cipher chosen for AES. The test data published by ETSI and 3GPP assume that Rijndael is used. Vendors are expected to support the use of Rijndael in these functions. Thus, the security of these functions will depend on Rijndael. Rijndael was designed by Joan Dæmen and Vincent Rijmen and was chosen to be the Advanced Encryption Standard (AES) in 2000 by the U.S. National Institute of Standards and Technology. The U.S. National Security Agency (NSA) still approves of the use of Rijndael to encrypt TOP SECRET information.

The most successful attack on Rijndael was published by Nicholas Courtois and Josef Peiprzyk in 2002 [Cou02]. This attack reduces the strength of 128-bit Rijndael to 91 bits—that is, about 10^{28} operations are required. This indicates that Rijndael will be secure for at least the next 10–15 years. If Rijndael is exploited, then the cryptographic primitive used by 3G can be changed, although this will require significant updates for the SIMs. Thus, the primitives used for key generation and for the generation of nonces can be considered secure.

The authentication string is 64 bits. If the authentication function is considered strong, this means that there is a 2^{-64} ($\sim 10^{-19}$) chance that a malicious user will be able to impersonate a user and steal sessions or gain access to other functions. Note that in the banking industry, PINs on ATM cards are four to six digits long, and three attempts are permitted. This means that there is a chance of at least 10^{-6} of accessing the account by guessing the PIN. Thus, the authentication using the SIM is significantly more secure than these cards. If the 64 bits of authentication were ever considered insufficient, then 3G includes a reauthentication that would be used to authenticate the SIM twice. Upon receipt of the first authentication, immediate re-authentication would be requested with another random challenge: that is to set the re-authentication period to be small on the first registration. This will allow the 128 bits of the shared authentication key to be used.

3.6.4.6 3 GSM Summary

The security model used by 3G is fundamentally sound, the main risk being fraud originating from a trusted domain. Carriers, therefore, must take steps, both technically and commercially, to ensure that their exposure to this risk is small.

The security protocols used by 3G assuming the use of a SIM have been analyzed and have been found to be sound. The cryptographic primitives are currently secure and are expected to be secure in the foreseeable future. The cryptographic primitives for authentication and key agreement are chosen by the carrier, although Rijndael will be supported by vendors. Therefore, if the cryptographic primitives for authentication and key agreement are compromised or are discovered to be weak, then the carrier is able chose other primitives.

If the confidentiality and integrity primitive, Kasumi, is compromised, then changes to the standards and equipment will be required. However, this will be a global problem affecting all 3G deployments; therefore, finding a solution will an industry problem and the expense is amortized over many carriers and users. However, it is very unlikely that in the next 10–15 years such a compromise will be published.

3.7   References

[Arba2001] W. Arbaugh, Your 802.11 Wireless Network has no Clothes. March 2001

[Biha2005] E. Biham, Dunkelman, N Keller, A Related-Key Rectangle Attack on the Full KASUMI, ASIACRYPT 2005, pp 443–461.

[Colb1997] B. Colbert, On the Security of Cryptographic Algorithms, PhD Thesis, University of UNSW, 9 July 1997

[Cour2002] N. Courtois, J. Pieprzyk, Cryptanalysis of Block Ciphers with Overdefined Systems of Equations, ASIACRYPT 2002, 8th International Conference on the Theory and Application of Cryptology and Information Security, Queenstown, New Zealand, December 1-5, 2002. pp 267-287.

[ISO 27002] ISO/IEC 27002. Information Technology – Security Techniques – Code of practice for information security management.

[IEEE 802.11i 2004] IEEE Standard for Information Technology – Telecommunications and information exchange between systems – Local and Metropolitan area networks – Specific Requirements. Part 11

Wireless LAN Medium Access Control (MAC) and Physical Layer (PHY) specifications Amendment 6: Medium Access Control (MAC) Security Enhancements. IEEE Computer Society.

[Køie2002] G.M. Køien, A Validation Model of the UMTS Authentication and Key Agreement Protocol, Technical Report R&D N 59/2002, Telnor, 20 December 2002

[Lehe2006] G. Lehembre, Seguridad WiFi- WEP, WPA,WPA2, hakin9 n°1/2006

[NIST SP800-48] National Institute of Standards and Technology. Special Publication. Wireless Network Security for IEEE 802.11a/b/g and Bluetooth

[NIST SP800-97] National Institute of Standards and Technology. Special Publication. Establishing Wireless Robust Security Networks: A Guide to IEEE 802.11i.

[OGC 2007] Office of Government Commerce, The Official Introduction to the ITIL Service Lifecycle, 2007

[OGC 2007 2] Office of Government Commerce, Service Transition, 2007

[RFC 1155] Internet Engineering Task Force, Structure and identification of management information for TCP/IP-based internets

[RFC 1157] Internet Engineering Task Force, Simple Network Management Protocol (SNMP)

[RFC 2865] Internet Engineering Task Force, Remote Authentication Dial In User Service (RADIUS)

[RFC 3584] Internet Engineering Task Force, Coexistence between Version 1, Version 2, and Version 3 of the Internet-standard Network Management Framework

[RFC 3588] Internet Engineering Task Force, Diameter Base Protocol

[RFC 3748] Internet Engineering Task Force, Extensible Authentication Protocol (EAP)

[TMF 2007] TeleManagement Forum, Enhanced Telecom Operations Model (eTOM) – The Business Process Framework, 2007

[TS 33.120] 3GPP Technical Specification Group Services and System Aspects; Security principles and objectives

[TS 33.105] 3GPP Technical Specification Group Services and System Aspects; Cryptographic algorithm requirements

[Zel99] D. Zeltserman, A Practical Guide to SNMPv3 and Network Management. Prentice Hall. 1999

Chapter 4

Radio Frequency Engineering, Propagation and Antennas

4.1 Introduction

A significant amount of time is spent in the design and development of front-end antennas and radio-frequency (RF) devices. Wireless engineers must be well versed in RF propagation issues if they are to factor in parameters for developing mitigation techniques. This chapter describes what a practicing engineer must know in the areas of antennas, propagation, and RF engineering.

4.2 Contents

This chapter covers the following topics:

Antennas
Propagation
RF engineering

4.3 Antennas

4.3.1 Background

Without an antenna, a wireless system cannot perform its function. From the wireless system's perspective, the primary use of the antenna is to propagate an information-carrying signal from the transmitter to the receiver. The antenna functions as a transducer. It converts a guided wave from a waveguide, coaxial cable, or transmission line into a propagating electromagnetic (EM) wave in free space (or air). The guided wave excites electrical currents that flow in the antenna structure, and these electrical currents in turn develop an electromagnetic field surrounding the antenna according to Faraday's Law and Ampere's Law. This EM field generates the wave that propagates through free space outward from the transmitting antenna toward the receiving antennas.

Antennas are typically reciprocal devices: they can receive and process EM waves while they transmit. The reverse process occurs when the antenna receives—the incoming EM wave excites electrical currents on the antenna structure. These currents are converted into guided waves that carry the information signal to the RF circuitry for processing and demodulation. In view of this dual capability of the antenna(transmit and receive), a separate device called the duplexer is attached to the antenna to separate the outgoing (transmit) and incoming (receive) signals. Two spatial regions surround an antenna: the *near field* and *far field*. The near-field region is closest to the antenna and contains the reactive and oscillating EM field. The far-field region is farther away, where only the transverse-propagating EM field exists. The distance to the far-field region is generally approximated as $R \geq 2D^2/\lambda$, where D is the largest dimension of the antenna and λ is the wavelength. The transition between near- and far-field regions is not abrupt but gradual, as the reactive near field sheds energy into the propagating wave. The far-field distance is generally a good approximation of the distance required to see only the outward propagating wave.

An antenna can also be considered as an impedance transformer because it converts a nominal RF system impedance (e.g., 50 ohms) into the free-space wave impedance (~377 ohms). Antennas generally operate over a narrow, or limited, bandwidth; hence, they can also be regarded as bandpass filters in the overall communication system. Depending on the antenna's bandwidth, a pre-selection filter in the receiver may not be needed. Antennas can also be designed for wideband or broadband operation.

4.3.2 Antenna Parameters

Any antenna for a wireless system cannot be thoroughly designed or analyzed without an understanding of its basic parameters. However, the relative importance of each parameter depends on the intended application. The following sections are adapted from reference [Kra02], and depend on IEEE Standard 145-83 [IEEE83] for some definitions.

4.3.2.1 Input Impedance

The antenna input impedance is measured at its feed terminals (i.e., where the guided wave structure is connected to the antenna). It is generally not 377 ohms but depends on the physical construction of the antenna and varies with frequency. Antenna input impedance is an important parameter because it generally must be matched to the characteristic impedance of the guided wave line and RF circuitry (e.g., 50 ohms) through an impedance transformer, or balun, at the feed terminals. Any large mismatch between the antenna impedance and the RF system impedance produces a large return loss and reduces the antenna gain and effective radiated power (ERP). This could degrade system performance to the point that the communication link would not close in applications where link margins are small. Most system designers require the antenna's return loss to be < 2:1 or lower over the bandwidth. The antenna structure, which may include the impedance-matching circuitry, often includes an RF connector that links it to the rest of the system. Antenna input impedance is determined either by closed-form solutions, through computer modeling and simulation, or by measurement with a network analyzer.

4.3.2.2 Size, Weight, and Power (SWAP)

For certain applications, size, weight, and power requirements are critical, particularly when the antenna is portable or mounted on a vehicle (aircraft, satellite, etc.). SWAP considerations are critical for satellite applications because space in the satellite and inside the launch vehicle is extremely limited. So is the power-generating capability on board a satellite and added weight is very costly to launch. However, SWAP may be of secondary importance for fixed metropolitan Wi-Fi installations where AC power is conveniently available and weight is not a major concern. The most extensively used antennas today are in cell phones that operate on a small battery. While initial versions of cell phones used a stub antenna whose size, weight, and power were of some concern, the trend has now moved towards hiding the antenna inside the plastic casing, making size, weight, and power extremely important for cell phone manufacturers. For a fixed operating frequency, the higher the gain required by the system, the larger the antenna (and hence weight). For a fixed physical size, operating at a higher frequency provides higher gain, while a lower frequency provides lower gain.

4.3.2.3 Field and Power Patterns

Virtually all antennas radiate electromagnetic waves more strongly in some directions than in others. This behavior is measured by using *field patterns*, which are three-dimensional (3D) vector quantities that are functions of θ and ϕ, the angles used in a standard spherical coordinate system. The field patterns describe the variation of the components of the electric (or magnetic) field vector as a function of θ and ϕ. For example, a normalized field pattern is defined by

$$E_\theta(\theta,\phi)_n = E_\theta(\theta,\phi)/E_\theta(\theta,\phi)_{max} \qquad (4.3.1)$$

and

$$E_\phi(\theta,\phi)_n = E_\phi(\theta,\phi)/E_\phi(\theta,\phi)_{max} \qquad (4.3.2)$$

Where E_θ and E_ϕ are the θ and ϕ components of the spherical electric-field vector. There is no radial component of the electric field, E_r, in the far field of the antenna due to the transverse nature of the electromagnetic waves. Field patterns are typically displayed in 3D color plots or by using several two-dimensional planar cuts. These 2D cuts, usually shown as polar plots, are often the principal planes (*xz* and *yz* planes). They are called the E-plane cuts, while the *xy* plane is known as the H-plane cut. *Power patterns* are visualized in 3D or in 2D planar cuts in exactly the same way as field patterns. Power patterns are defined by

$$P_n(\theta,\phi)_n = S(\theta,\phi)/S(\theta,\phi)_{max} \qquad (4.3.3)$$

where S is the Poynting vector given by

$$S(\theta,\phi) = \left[E_\theta^2(\theta,\phi) + E_\phi^2(\theta,\phi)\right]/Z_0 \qquad (4.3.4)$$

and Z_0 is the free-space wave impedance (376.7 ohms). Field and/or power patterns for an antenna are determined either from closed-form solutions, from computer modeling and simulation, or from antenna measurements.

4.3.3 Beamwidth

One of the most important antenna parameters is beamwidth. An antenna that is directive and radiates its power more strongly in one direction has one major (or main) lobe, also called the *main beam*, in its field (or power) pattern and several minor lobes, or *sidelobes*. The beamwidth (in degrees) is a specification of the antenna's main beam; it is a measure of how strongly the antenna radiates in the main beam's direction. For example, a half-wave dipole antenna has a beamwidth of 78°, but a large parabolic reflector antenna may have a beamwidth of only 1°. Thus, the parabolic reflector is more highly *directive* than the dipole because it radiates more strongly along its main beam. The lower the beamwidth of the antenna, the more directive it is and thus it has higher gain. Beamwidth is inversely proportional to antenna gain, and thus is inversely proportional to antenna size. For a fixed physical antenna size, increasing the operating frequency increases the gain and therefore decreases the beamwidth. Beamwidth is most often expressed as the -3 dB (or half-power) beamwidth, which is defined where the field pattern

$$E_\theta(\theta,\phi)_n = 1/\sqrt{2} \qquad (4.3.5)$$

This point corresponds to the angle between where the power pattern

$$P_n(\theta,\phi)_n = 1/2 \qquad\qquad (4.3.6)$$

because power is related to E_θ^2, and -3 dB = 10log(0.5). Other beamwidth specifications may also be used such as -10 dB or -20 dB beamwidth, which correspond to where $P_n(\theta,\phi)_n = 0.1$ and $P_n(\theta,\phi)_n = 0.01$, respectively. Half-power beamwidth is also expressed in the E- and H-planes as θ_{HP} and ϕ_{HP}, where θ_{HP} is defined in the *xz* or *yz* plane and ϕ_{HP} is defined in the *xy* plane.

4.3.4 Directivity, Gain, and Aperture

The *directivity, D,* of an antenna is defined as the ratio of the radiation intensity to the radiation intensity averaged over a sphere. Directivity is a function of θ and ϕ and is mathematically expressed by

$$D(\theta,\phi) = U(\theta,\phi)/U(\theta,\phi)_{avg} = S(\theta,\phi)/S(\theta,\phi)_{avg} \qquad\qquad (4.3.7)$$

where $U(\theta,\phi)$ is the radiation intensity. If no direction is specified, the direction of maximum radiation intensity is implied and the directivity becomes a single value. The gain, *G*, of an antenna is less than the directivity due to ohmic and other losses. The gain is expressed mathematically as $G = kD$, where *k* is an efficiency factor between 0 and 1.

An *isotropic radiator* (or isotropic point source) is a theoretical antenna that has an omnidirectional pattern and a directivity $D = 1$. This means that the isotropic source radiates energy uniformly in all directions over 4π steradians. All antennas have directivity $D \geq 1$. Gain is usually specified in units of dBi, which means dB over isotropic. Thus, an isotropic radiator has a gain of 0 dBi. For example, a half-wave dipole has a directivity $D = 1.64$, which is equivalent to a gain of 2.15 dBi, and a gain pattern that is approximately $sin^3\ (\theta)$. The 3D gain pattern of a half-wave dipole can be visualized as a doughnut, with the dipole antenna positioned vertically at the doughnut's center. The pattern has a half-power beamwidth (HPBW) of 78° in the E-plane (elevation plane) and an omnidirectional (uniform) pattern in the H-plane (azimuth plane).

Although mathematically different, the terms gain and directivity are often used interchangeably. Most antenna engineers plot 2D and 3D directivity or gain patterns in place of field or power patterns. This is because the gain or directivity pattern is very similar in definition to the normalized power pattern, and the gain is often the single most important parameter of the antenna. In any wireless communication system, the antenna must have enough gain to close the link and perhaps provide some link margin as well. An approximate expression for directivity is given by $D = 40,000/(\theta_{HP}\phi_{HP})$, where θ_{HP} and ϕ_{HP} are the E- and H-plane half-power beamwidths in degrees. The directivity and gain of an antenna can be determined either from closed-form solutions, from computer modeling and simulation, or from measurements.

Every antenna has an *effective aperture,* or A_e. Effective aperture can be viewed as that area over which the antenna effectively "captures" incoming electromagnetic waves. This aperture generally differs from the antenna's physical aperture, which is the cross-sectional area of the antenna's physical structure. Thus, the effective aperture is essentially an "electrical area" of the antenna representing its electromagnetic collection area. The effective aperture can be larger or smaller than the physical aperture, depending on the antenna type. The directivity (and hence gain) of an antenna is related to the effective aperture by the expression

$$D = 4\pi A_e / \lambda^2 \qquad (4.3.8)$$

Therefore, more gain requires either a larger effective aperture or area, or operation at a shorter wavelength (i.e., a higher frequency).

4.3.5 Polarization

The polarization of an antenna describes the orientation of the electric-field vector of the radiated electromagnetic field in the antenna's far field as the field propagates away from the antenna. Antennas radiate linear, elliptical, or circular polarization. In linear polarization (LP), the electric-field vector oscillates along a straight line as the wave propagates outward. Examples include antennas that are vertically polarized, horizontally polarized, or perhaps have a linear 45° polarization. Dipole or monopole antennas are good examples of linearly polarized antennas. An automobile radio antenna is linearly polarized. An antenna designed for vertical polarization will not receive or transmit signals that are horizontally polarized. In this case, the horizontally polarized signal represents the *cross-polarization,* or *cross-pol,* term.

For circular polarization, the electric-field vector rotates and traces out the locus of a circle as the radiated wave propagates. Circular polarization (CP) is further divided into left-hand circular polarization (LHCP) and right-hand circular polarization (RHCP), depending on the sense of rotation. With the thumb of the right hand pointed in the direction of propagation, curling the fingers of the right hand give the sense of RHCP. Thus, looking in the direction of propagation, the electric-field vector rotates clockwise for RHCP and counterclockwise for LHCP. Antennas designed for RHCP will not receive LHCP and vice versa. LHCP represents the cross-pol for RHCP and vice versa. A helical antenna is a prime example of a circularly polarized antenna.

With elliptical polarization (EP), the electric-field vector traces out the locus of an ellipse as the radiated wave propagate. Elliptical polarization can also be given a RH or LH sense in the same way as circular polarization. In fact, both linear and circular polarizations are a special case of elliptical polarization. A measure of an antenna's polarization is its *axial ratio (AR)*, which is the ratio of the major to minor axes in the polarization ellipse. For linear polarization, $AR = \infty$, and for circular polarization, $AR = 1$.

Polarization mismatch can be a significant concern in wireless system design, implementation, and operation. For example, a linearly polarized antenna can receive CP signals, but with 3 dB less power due to the polarization mismatch. Similarly, a CP antenna can receive LP signals, but again with the 3-dB loss. Therefore, any sound link-budget analysis must always include a loss term for polarization mismatch, depending on the system application. For maximum power transfer and signal reception, the transmitter (Tx) antenna must be of the same polarization and orientation as the receiver (Rx) antenna. If the Tx is vertical LP, the Rx must also be LP and oriented vertically.

4.3.6 Antenna Types

Antennas can be classified in many ways—by their frequency bandwidth (narrowband, broadband, ultrawideband), by their physical design (e.g., wire antennas, reflectors, printed circuit, aperture), by their pattern characteristics (omnidirectional, high gain), by their electrical size (small, large), by their operational mode (traveling wave, surface wave, guided wave), or by their scanning characteristics (none, mechanical, electrical). A single antenna design may span several categories but perhaps the most straightforward way to classify antennas is by their physical design.

Wire antenna designs may involve loops, dipoles, folded dipoles, rhomboids, long wires, twin lines, and helices, and may include phased arrays of these individual elements. Many wire antennas are used in applications where lower gain (2–3 dBi), linear polarization, and omnidirectional patterns are acceptable. Applications include portable radios, automobiles, mobile handsets, and wireless Internet modems and routers. Some wire designs such as helix antennas can have more moderate gains in the 8 dBi to 15 dBi range. Wire antennas are typically narrowband antennas with bandwidths on the order of 10%–20% due to their resonant behavior. Other types of wire-like antennas such as Yagi-Udas, log periodics, and conical spirals operate over a much broader bandwidth than dipoles or other wire antennas. Log-periodic antennas are a mainstay for receiving broadcast television signals.

Aperture antennas consist of reflectors, parabolic dishes, lenses, spirals, flat panels, frequency-selective surfaces (FSS), and horns that radiate electromagnetic energy from a physical aperture. Such antennas are used in applications calling for moderate gains ($\approx$ 3 dBi to 20 dBi) or high gains (> 20 dBi) with moderate to very narrow beamwidths. Applications for aperture antennas include satellite communications, radar, antenna measurements, and microwave telephone relays. An in-depth discussion of each antenna type is beyond the scope of this chapter, but any good undergraduate or graduate-level antenna textbook provides detailed descriptions. An exhaustive reference is the *Antenna Engineering Handbook* [Vol07].

Antennas typically operate over a narrow bandwidth (< 20%) due to their high Q structure, abrupt transitions, and resonant behavior. However, they can be designed to operate over moderate-to-large bandwidths by ensuring, for example, smoother variations and transitions in the antenna structure. Broadband antennas are often called frequency-independent antennas, and can exhibit moderate-to-large bandwidths, anywhere from 2:1 up to 33:1 or more. Rumsey's principle states that the impedance and pattern properties of an antenna will be frequency-independent if the antenna shape is specified only in terms of angles [Kra02]. Examples of these broadband types are the biconical, planar log-spiral, and conical log-spiral antennas. A broadband antenna may also combine several different elements or regions of varying size or length so that the active region of the antenna shifts as the frequency changes. An excellent example of this type is the log-periodic antenna.

Aperture antennas can also be made broadband by including structures with smooth transitions and variations such as the quad-ridge horn antenna or impulse-radiating antenna. A new class of broadband antennas called fractal antennas are being developed and deployed based on self-similar geometrical structures [Wer00].

4.3.7 Phased Arrays

There are many wireless applications where the antenna pattern must be reconfigured or the antenna beam moved to different angular locations to communicate with different end users or to scan the surrounding area. These applications are addressed by moving (i.e., steering or scanning) the antenna beam either mechanically or electrically. Mechanically scanned antennas are typically some form of aperture antenna mounted on a pedestal that rotates in the azimuth direction, and they often include some rotation or movement in the elevation direction. Antennas for radars, satellite communication, and radio astronomy are examples of mechanically scanned designs. Electrically steered antennas are called *phased array antennas*, or simply *phased arrays*. A phased array antenna generally remains fixed in space while the antenna beam is scanned (or even reconfigured) electronically. However, some applications may use a mechanically scanned phased array. In this case, the mechanical scan may cover the azimuth while the phased array scans in elevation.

A phased array antenna is composed of identical antenna elements in a regularly spaced lattice of some kind. A linear phased array places the antenna elements at equal intervals along a straight line, usually ¼ to ½ wavelength apart. A linear phased array can scan only along the direction of the array. A flat-panel (or 2D) phased array places the antenna elements at equal *x-y* space intervals on a planar surface. A flat-panel phased array has two scanning degrees of freedom.

The operation of a phased array is based on the principle of constructive or destructive interference. Signals feeding each element are given different amplitude weights (i.e., an *amplitude taper*) and relative phases (i.e., a *phase taper*). Signals from each element traveling to the same point in the far field of the array experience different phase changes due to the distance traveled; combined with the relative signal phases, they add or subtract when summed together. The amplitude taper and the interelement spacing control the sidelobe levels and the phase taper controls the position of the main beam. Thus, the beam is electronically scanned by changing the relative phase of the signal feeding each antenna element.

Phase shifters are usually digitally controlled, and it is important to understand the relationship between a one-bit change in phase control to the amount of change in scan angle. The interelement spacing between elements in the phased array is an important design parameter. A spacing greater than one wavelength introduces *grating lobes*, unwanted sidelobes that can reach levels almost equal to the main beam. This phenomenon degrades system performance by transmitting or receiving signals in unwanted directions.

The *array factor* for a phased array is the antenna pattern produced using a chosen amplitude and phase taper by assuming each antenna element is a point source. The amplitude, interelement spacing, and phase taper are iteratively adjusted by examining the array factor until the desired sidelobe-level performance is achieved. If the antenna elements in an array are identical, then the composite phased array antenna pattern is found simply by multiplying the array factor by the individual element antenna pattern. If the array elements are not identical, the composite array pattern is obtained by a vector sum in the far-field region of the array of the electromagnetic field radiated from each array element.

4.3.8 Beamforming and Smart Antennas

A traditional phased array antenna has a pattern with a single main beam and various sidelobes. The angular position of the beam can be electronically steered by varying the phase of the phase shifter at each antenna element. The level of the sidelobes is controlled by varying the amplitude of the signal to each element and the physical spacing between elements. A phased array antenna can also produce multiple beams by using a *beamforming* network behind the antenna. This network is a complex interconnection of phase shifters, directional couplers, power dividers or combiners, or transmission lines that feed each of the array elements. Many different types of beamforming networks exist, including a power divider, a Butler matrix, Blass and Nolen matrices, a Wullenweber array, a McFarland 2D matrix, a Rotman lens, a Bootlace lens, and a dome lens [Han98]. These beamforming techniques produce multiple simultaneous fixed beams, and an in-depth discussion is beyond the scope of this chapter. Multiple independently steerable beams can be produced by having more than one set of phase shifters and amplitude-control elements connected to the array elements.

Equivalently, multiple independent beams can also be produced with a *subarray*. This is simply a set of phased array elements that is part of a larger phased array system. For example, a 32×32-element phased array antenna could be spatially subdivided into four 8×8-element subarrays to provide four independently steerable beams. But, the subarray elements need not necessarily be grouped together spatially. The 32×32 array could be subdivided according to the rule that every fourth element belongs to one of the subarrays. In this way, the subarrays are interdigitated with each other and should produce independent beams provided the subarrays have sufficient electrical isolation.

Smart antennas incorporate beam-steered phased arrays and use signal-processing techniques to shape the beam pattern according to certain optimum criteria [Vol07]. For example, a smart antenna can create a null in the antenna pattern in the direction of an interfering source or jammer. Smart antennas can increase the capacity of wireless communication networks through space division multiple access (SDMA). Some of the basic signal-processing algorithms used in smart antennas are the least mean squares (LMS), sample matrix inversion (SMI), recursive least squares (RLS), conjugate gradient method (CGM), constant modulus algorithm (CMA), and the least-squares constant-modulus algorithm (LS-CMA) [Vol07]. Smart antennas are also often called digital beamformed or adaptive arrays.

4.3.9 Antenna Design and Measurements

To begin an antenna design, the requirements must first be defined. These requirements typically flow from a rigorous system-level design and analysis. Parameters such as gain, return loss, beamwidth, axial ratio, polarization, and bandwidth must be well-defined and realistic. Mechanical considerations of the system may indicate the type of basic antenna structure that best suits the application. For example, an aircraft antenna might need to conform to the aircraft surface; thus, a spiral or patch antenna may work best. Onboard a satellite, a helix antenna, or reflector antenna may be more suitable.

The design can proceed once the antenna requirements are defined. An antenna can be designed in several ways depending on its basic structure. For very simple antennas, a closed-form solution may be available. Often, a set of parametric curves can be used to set the various physical antenna parameters to achieve the desired gain, bandwidth, etc. Another popular design approach is to use commercial electromagnetics software to design the antenna and analyze its performance by computing the gain, bandwidth, return loss, axial ratio, and beamwidth. These commercial suites are based on four primary numerical solution techniques for Maxwell's equations—the finite element method (FEM), the finite-difference time-domain (FDTD) method, the method of moments (MoM), and physical optics (PO). The High Frequency Structure Simulator (HFSS) from Ansoft Corp. uses FEM, and Microwave Studio from Computer Simulation Technology Inc. (CST) uses an FDTD-like technique. The Numerical Electromagnetics Code (NEC) is a MoM software tool, and the GRASP tool, which uses PO, is produced by TICRA. GRASP is used almost exclusively for reflector antenna design and analysis, particularly for large reflectors.

The design process is usually iterative, starting with a basic design followed by either fabricating and testing the antenna to gauge its performance, or by performing several design iterations through computer analysis and simulation. The design is modified after each iteration. Design by computer is generally preferable because it eliminates expensive fabrication and testing. Once the final design is chosen, a prototype can be fabricated and experimentally characterized.

Antenna gain, patterns, and effective radiated power (ERP) are determined in one of two ways: by computer simulations or experimentally through measurements. Often, both techniques are used, with measurements generally viewed as the "right" answer to validate the computer analysis. Antenna measurements simulate how the actual physical device will operate in practice, provided the measurement process is sound. Therefore, they take precedence in verifying the final antenna performance.

According to IEEE Std 145-1983, ERP is defined as: *in a given direction, the relative gain of a transmitting antenna with respect to the maximum directivity of a half-wave dipole multiplied by the net power accepted by the antenna from the connected transmitter* [IEEE83]. From the same standard, the effective isotropic radiated power (EIRP) is defined as: *in a given direction, the gain of a transmitting antenna multiplied by the net power accepted by the antenna from the connected transmitter* [IEEE83]. ERP and EIRP are often confused in practice. EIRP is generally used in satellite-link budget calculations for RF system design and analysis.

Antenna gain, patterns, and impedance (or return loss) are determined experimentally through antenna measurements. Impedance (or return loss) involves a simple measurement with a network analyzer. Gain patterns are measured by far-field or near-field procedures. Far-field measurements are done on a far-field antenna range. Such ranges are usually one of two types: an outdoor open area test site (OATS) or an indoor compact range/anechoic chamber. An OATS usually consists of two towers or other outdoor locations separated by line-of-sight over a fixed distance. One tower contains the transmitting antenna along with the measurement equipment such as a network analyzer and antenna positioner control. The other tower holds the antenna under test (AUT). It may contain a positioner to scan the AUT in azimuth and/or elevation and to change orientation for cross-pol measurements.

The indoor anechoic chamber, on the other hand, is usually a moderate-to-large room electromagnetically shielded from outside stray EM signals. Electromagnetic absorber material is placed on the ceiling, walls and floor of the chamber to reduce or eliminate reflections of test signals that could corrupt the measurements. The chamber uses a transmit antenna and the AUT is the receive antenna. A compact range is a smaller version of the anechoic chamber and uses a reflector to generate a plane wave that impinges on the AUT. The transmit antenna for the compact range is typically a calibrated horn antenna that is used as a feed to illuminate the reflector, usually in an offset configuration.

In a far-field measurement, the antenna measurement range and equipment are first calibrated. Typically, the range is calibrated using standard-gain horn antennas because the gain and patterns of these antennas are very well known. Then the AUT is mounted on a positioner that mechanically moves the antenna in azimuth and often in elevation. The positioner can also rotate the antenna to provide measurements of both polarizations. Both the magnitude (I) and the phase (Q) of the received signal are recorded and are often transformed into antenna gain by the measurement control software, which is used to automate the measurement procedure.

In a planar near-field measurement, the AUT transmits while a small waveguide probe or horn antenna is mechanically scanned across a planar aperture in front of the AUT. Both I and Q of the electromagnetic field are recorded at each sample point, and the sampling density must meet or exceed the Nyquist criterion. Control of the longitudinal probe position is critical to achieve accurate and reliable results. The I and Q data are then transformed to the far field in software by aperture integration to obtain the far-field pattern. The main advantage of planar near-field scanning is its mathematical simplicity because the software transformation can employ an FFT algorithm. The main disadvantage is that the far-field pattern is accurate only over a limited angular range. Planar near-field measurements are best suited to high-gain antennas.

Cylindrical and spherical near-field scanning are also available, which scan the probe over a cylindrical or spherical surface surrounding the antenna. A similar (but more complicated) aperture-integration technique is used on the sample data to compute the far-field pattern. Spherical near-field scanning works best for low-gain or more omnidirectional antennas. But for any near-field scanning technique, the equipment is also initially calibrated, and the final data must be modified to compensate for the field pattern of the probe.

4.3.10 RF Site Surveys

In both the planning and operational phases of deploying an indoor or outdoor wireless network, RF site surveys must be performed to optimize network performance and to reduce implementation, operating, and maintenance costs. For an indoor wireless network, the basic steps are to obtain a facility diagram (blueprint, etc.), visually inspect the facility, identify user areas, determine preliminary access-point locations, verify final access-point locations, and document the findings [Gei02]. Often, an RF site survey involves walking around the facility and measuring the signal strength either from wireless network access points or from potential sources of interference (e.g., microwave ovens) or from both. Approaches vary from the simplest case of walking around with just a signal strength meter or a laptop with a wireless network-interface card (NIC) and specialized RF signal-mapping software, all the way to a sophisticated setup with GPS location recording and a spectrum analyzer.

In any case, network performance will be best if a walkaround survey is completed with a spectrum analyzer to identify possible interference sources at the start of the design phase. Software planning tools exist that allow the user to create or import building layouts and plan the wireless network layout using RF propagation simulators. These tools reduce testing costs by identifying initial locations of access points that can then be fine-tuned with fewer walkaround surveys. A number of manufacturers offer special testing equipment such as spectrum analyzers, Wi-Fi analyzers, stimulus transmitters, power meters, and receivers.

Outdoor surveys for wireless cellular networks are done in much the same way as indoor surveys, but they deal instead with terrain maps rather than building plans. Signal-strength measurements are done with RF drive surveys in a vehicle outfitted with antennas and measurement equipment. Software planning and analysis tools also exist for cellular networks to analyze multipath propagation and signal strengths for a particular coverage area. Much of the measurement equipment is similar to that for indoor networks, though more specialized. The basic measurement procedures are also similar. They identify potential sources of interference, record signal strengths at various locations and then optimize the network coverage and performance. Site surveys must be done periodically as part of an ongoing network operations plan.

4.3.11 Diversity

In any wireless communication system, transmitted signals propagate over a variety of paths to the receiving antenna due to scattering and diffraction by manufactured and natural objects such as the ground, buildings, vehicles, trees, hills, mountains, and other structures. This phenomenon is known as *multi-path propagation*. These propagation paths can add attenuation, distortion, delays, de-polarization, or phase changes to the signals. They can also be time-varying when communicating with a mobile phone in a vehicle. When these signals are coherently summed in the receiver, constructive or destructive interference occurs [Hou04]. With destructive interference, the power of the resulting signal can be significantly reduced. This phenomenon of destructive interference is called *fading*. *Deep fading* occurs when the combined signals are almost 180° out of phase. Fading can severely impact the performance of a wireless system [Vol07, Rap02].

To combat fading, two or more copies of the transmitted signal can be combined at the receiver to increase the signal power. The basic premise is that while some transmitted signals may experience fading, others may not. By using several copies of the signal and combining them at the receiver, the signal power can be increased and system performance enhanced. This is the basic concept of *diversity*. Diversity takes advantage of the statistical properties of the time-varying communication channel by assuming that the signals combined at the receiver experience fading independent of one another.

There are four primary types of diversity: frequency, time, space, and polarization. In frequency diversity, the information signal is modulated onto several different carrier frequencies, each separated by at least the coherence bandwidth of the fading channel, so that each signal will fade independently of the others. With time diversity, the desired signal is transmitted in several different time intervals separated by at least the coherence time of the fading channel. Again, this is done so that each signal will experience independent fading. Space diversity involves receiving the desired signal using several antennas, each separated in space at the receiver. The idea here is that signals reaching different antennas will fade independently, and so could be combined to maximize signal power. Spatial diversity can be effectively equivalent to a phased array, depending how the received signals are combined at the receiver [Die00, Hou04, Rap02, Vol07].

Polarization diversity involves transmitting the desired signal using two orthogonal polarizations because the polarizations have very low correlation. Thus, it is unlikely that both communication channels will experience a deep fade simultaneously. Therefore, the received signals could be combined to maximize received signal power. Polarization diversity could effectively double the network capacity in a robust implementation [Hou04, Die00]. For optimal performance, there has to be sufficient electrical isolation between the two polarizations in both the transmit and receive antennas.

There are three basic techniques for diversity combining at the receiver: selection diversity, equal-gain combining, and maximum-ratio combining. In selection diversity, the signal with the highest received level is switched into the receiver. In equal-gain combining, all received signals are coherently summed with equal amplitude and phase. In maximum-ratio combining, a weighted summation of the signals is performed where the amplitudes are proportional to the signal-to-noise ratio (SNR) for each signal and the phases are kept equal. When they are used with spatially separated antennas, equal-gain and maximum-ratio combining result in phased array antennas. The measure of diversity performance is the *diversity gain*, which is the improvement in signal level when diversity techniques are used versus the signal level when they are not [Die00].

There are three basic types of antenna diversity: space, pattern (or angle), and polarization. Space and polarization diversity were mentioned previously. Pattern (or angle) diversity involves discriminating among the propagation channels based on angle. For example, it could mean using antennas with different or orthogonal radiation patterns or using a multiple-beam antenna to receive the multipath signals that arrive from different angles [Dit00, Hea08].

4.3.12 System Capacity and Multiple Access

The capacity of any wireless communication system is often most limited by the frequency spectrum. Capacity improvements generally focus on frequency *reuse* and more efficient use of the spectrum. Reuse refers to the ability to use some system resource to create multiple physical channels that occupy the same frequency spectrum. System capacity can be increased provided the phenomenon of *co-channel interference* (CCI) can be mitigated. CCI occurs when two transmitters transmit on the same frequency channel. If the two signal levels at a receiver are sufficiently high, the receiver cannot distinguish between the two signals, causing CCI. On the other hand, CCI will not occur if one signal level is much lower than the other.

The spectrum allocated for a wireless system is typically divided into many different frequency channels to allow multiple users to communicate. This concept is called *multiple access*. The diversity concept should not be confused with multiple access or reuse. Frequency reuse for cellular telephone systems is accomplished through spatial separation by using the same frequency in different geographical areas, or cells. Cells are separated by enough distance so that CCI is below a certain threshold. In spatial division multiple access (SDMA), frequencies can also be reused by using directional antennas. In this case, to avoid CCI the antennas should have low sidelobe levels and low front-to-back ratios [Die00].

System capacity can be increased through *cell splitting* and *sectoring*. Cell splitting subdivides a crowded cell into smaller cells where each new cell has its own base station with an antenna but with lower transmit power. Capacity is increased because the number of channels per unit area has increased. In sectoring, the base station uses several directional antennas that each transmit within one angular sector. Typically, a cell is partitioned into three 120° or six 60° sectors. Directional antennas help reduce CCI and hence increase the signal-to-interference ratio (SIR). The reduction in interference leads to an increase in system capacity. However, in a sectored system, the antennas must be designed with the correct beamwidth in the azimuth direction to for the implementation to be successful [Rap02].

The issue with antennas and multiple access is that the antenna must transmit or receive over the entire frequency band spanning all the frequency channels of the wireless system. Each frequency channel does not have a separate antenna. Since antennas are generally resonant structures with narrow bandwidths, they must be selected or designed with enough bandwidth to cover the relevant frequency spectrum. System capacity can be improved through multiple access by decreasing the bandwidth of each channel and/or increasing the number of users per channel.

4.3.13 Base Station, Indoor, and Mobile Handset Antennas

Cellular base stations primarily use two types of antennas: omnidirectional and directional (or sectored). Omnidirectional antennas are typically used in rural areas or at lower capacity sites where sectoring is not required. Most urban or suburban cellular base stations use sectored antennas that must meet large capacity demands, with the 120° sector generally the most popular [And07]. Base station antennas are often mounted on stand-alone towers, but they are sometimes mounted on buildings, water towers, power line towers, and other manmade structures. Occasionally, the base station tower is camouflaged to look like a tree, flagpole, or other object that blends with the landscape.

Base station antennas are usually designed with moderate gains of anywhere from about 5 dBi to 24 dBi. Manufacturers will sometimes specify the antenna gain as dBd, which is the gain referenced to a ½-wave dipole. Recall that the ½-wave dipole has a gain of ≈ 2.1 dBi, and thus the gain in dBd will be 2.1 dB lower than the same specification in dBi. Base station antennas are typically designed for linear polarization (horizontal or vertical) and to radiate broadside, or at a 90° angle to the antenna structure. Therefore, when the antenna is mounted vertically on the tower, the main beam points in the horizontal direction. Some manufacturers offer dual polarized antennas (i.e., ± 45° polarization) to take advantage of polarization-diversity combining [And07]. Parabolic reflector antennas are used, particularly in the 2.4-GHz band for higher gain (≈ 20–24 dBi) outdoor wireless LAN applications. Flat panel antennas with ≈ 15–20 dBi gain are also used for both outdoor WLAN applications and to solve some indoor WLAN issues. Sectorized cellular tower antennas are often tilted downward so that the main beam points below the horizontal.

Careful attention must be paid to the mechanical design of the antenna and its enclosure, which must be protected from snow, rain, ice, and sun. Often this is accomplished using an enclosure known as a radome that is typically constructed of a lightweight, RF-transparent, robust material such as fiberglass. Furthermore, the antenna and its mount must be rigid enough to withstand severe wind loads that might otherwise mechanically move the antenna and result in a network outage (loss of signal) or increased interference. In addition, tower antennas must be protected from lightning strikes that could destroy the antenna and transmitter. Chapter 5 discusses these issues in more detail.

Indoor antennas are omnidirectional or directional and are used in such devices as access points and wireless routers for wireless local area networks (WLANs). Omni antennas provide the widest coverage and allow the system to be designed with circular overlapping cells from multiple-access points throughout the building. Indoor omni antennas generally have low gains of up to about 6 dBi, but using higher gain omnis can reduce the number of access points required, hence reducing system cost. Higher gain directional antennas (6 dBi to 20 dBi) can be used to cover large, narrow areas or in short links between buildings. Directional indoor antennas can also reduce the number of access points required in some applications. Manufacturers are continually developing new antennas with diversity to help solve building coverage issues in WLAN applications.

Designing an integral handset antenna is complicated due to size constraints and overall system issues. For handsets, antennas are typically low-gain and omnidirectional due to the random orientation of the handset at any given time. Furthermore, the handset antenna must be small, and small antennas are notoriously inefficient radiators. A retractable monopole is a good choice for a handset antenna, of which a portion may be wound into a coil for inductive loading. Other popular designs are based on a normal-mode helical antenna (NMHA) where a zigzag design is a popular option. This design can be easily printed on dielectric film and rolled into a round shape with a dielectric cover to form a short antenna stub for handsets.

The planar inverted F antenna (PIFA) is another good design to enclose in handsets; it is a combination of wire- and patch-antenna structures. It can also be designed for dual or tri-band use in global roaming handsets. Also in use are ceramic-chip antennas that more tightly confine the electromagnetic fields at the expense of much narrower bandwidth. Dielectric resonator antennas (DRAs) have also been designed. They work by exciting a mode inside a dielectric structure attached to the antenna. An example here is a dielectrically loaded, balanced, quadrifilar helix antenna. Finally, ferrites have been used in some applications to reduce the size of monopoles and other antennas [Vol07].

4.3.14 Antenna Tilt

The mechanical tilt angle of a base-station antenna is a well-known additional degree of freedom used to optimize radio networks. Since most of these antennas are installed in a vertical orientation and since their main beam is in the broadside direction (i.e., horizontal), mechanically tilting the antenna moves its main beam below the horizontal plane. The advantage of tilting is that it reduces path loss (PL), delay spread, transmitted power, and system interference. The tilt also strengthens signals from the home cell and reduces the CCI, thus increasing system capacity and improving network performance.

Antenna tilt is also an important specification used to develop the network topology in the planning phase. Early network deployments used a static tilt model where the tilt angle was preset and optimized during initial network operation. Any further changes in the tilt angle required a costly site visit by tower crews to re-position the antenna. More recent developments have focused on dynamic antenna tilt control [Cal06, Sio05, Pet04]. Base-station antenna manufacturers now offer remote electrical tilt (RET) modules built into the mounting structure to provide remote tilt control from the operations center. Antenna tilt can also be controlled remotely through the Internet via secured channels [Anr07]. Furthermore, engineers can dynamically control antenna tilt to optimize network performance depending on traffic characteristics. For example, preset tilt angles could be used at different times of day to accommodate varying traffic loads (e.g., rush hour, overnight). The RET uses a stepper motor to mechanically tilt the antenna to the desired angle. A phased array base-station antenna could be used equally well, with the main beam tilted electronically by adjusting the phase of the individual array elements. Care must be taken to insure that the number of bits used to quantize the phase shifters provides enough angular resolution for tilting the main beam.

4.3.15 MIMO

More recently, *multiple input/multiple output* (or MIMO) techniques have been applied that place multiple antennas at the transmitter and receiver to improve wireless system capacity and performance.

The basic idea behind MIMO is to exploit the space resource of the propagation channel and combine it with sophisticated signal processing to achieve significant gains in spectral efficiency. A MIMO demonstration by Lucent Technologies with 8 transmit and 12 receive antennas achieved a 1.2-Mb/s data rate in 30 kHz of bandwidth. A conventional radio link is single input/single output (SISO) and is subject to fading. A single input/multiple output (SIMO) channel offers diversity options at the receiver, such as space diversity, with multiple antennas. A multiple input/single output (MISO) channel offers diversity options at the transmitter, such as pattern or angle diversity. For example, a transmit phased array antenna can steer the main beam toward a particular receiver. Also, beamforming techniques can be employed at the transmitter to generate multiple beams.

Finally, for improved data rates a MIMO channel allows multiple data streams to be transmitted and recombined at multiple receive antennas. The system capacity improvement for a MIMO system is linearly proportional to the number of antennas [Con04]. However, in a MIMO system, many handheld receivers do not have multiple antennas. In such cases, maximum gain is achieved through transmit diversity by transmitting the same signal on multiple antennas. The premise is that one of the propagation channels to the receiver will likely have less fading (or none at all) compared to the other channels. When the receiver does have multiple antennas (true MIMO), maximum capacity is achieved through parallel transmission of different data streams. A good analogy for true MIMO is parallel computing where different portions of a large numerical simulation are sent to different processors, achieving much higher compute performance than a single processor. A current area of research in wireless communications is multiple-user MIMO (MU-MIMO) where the basic MIMO concept is extended to multiple receivers, each with multiple antennas [Ges03].

Manufacturers are also deploying aggressive beamforming technology, particularly for urban Wi-Fi networks to improve coverage and capacity over current configurations. The idea is to use multiple transmit antennas with digital beamforming technology to improve the wireless signal quality by maximizing antenna gain in the link. Essentially, this is an application of transmit diversity in a MISO link with a single Wi-Fi client. Through digital beamforming, interference is significantly reduced by focusing the antennas on a particular Wi-Fi client. SDMA can also improve wireless network capacity by leveraging the beamforming technology to enable simultaneous downlink channels with multiple users.

The U.S. Federal Communications Commission (FCC) recently encouraged the use of beamforming and SDMA techniques. In April 2005, the FCC revised Title 47, Part 15 of the U.S. Code of Federal Regulations that allowed increased directed power if beamforming and SDMA techniques are used. Conventional access points are limited to 36 dBm of EIRP (30 dBm of radiated power plus 6 dBi of antenna gain), but the new FCC rules allow that power to be increased if the total transmitted power is reduced by 1 dB for each 3 dB of antenna gain above 6 dBi [FCC08].

4.4   Propagation

4.4.1 Background

In mobile communications, the channel is the limiting factor in controlling system performance. Radio waves from transmitter to receiver can either have line-of-sight (LOS) propagation, or the path can be obstructed by buildings, trees, undulations of the ground, and other natural or man-made objects. Reflections, diffraction, and scattering are all factors that influence the link design. The complexity of the path is a function of the surroundings, which can either be urban, suburban, or rural. It is possible to have LOS between transmitter and receiver in a rural area, but generally impossible in urban and suburban settings. Thus, in general, the modeling process is based on statistics and measured data. A deterministic approach based on an assumption of LOS is only the starting point.

The propagation model in general predicts the average received level (large-scale variation) for an assumed separation between the transmitter (Tx) and the receiver (Rx), as well as the variability of the signal in an area surrounding the receiver (small-scale variation). The variability is the result of multipath, where the signal level can change by as much as 30–40 dB within a fraction of a wavelength. This multipath phenomenon follows a Rayleigh distribution.

Propagation inside a building, tunnel, or other enclosed area is no different than outside, where a typical receive signal is the composite sum of many signals reflected from walls, floors, ceilings, and other fixed objects. The modeling process is similar to propagation outside a building and is based on statistics or on empirical equations derived from measured data.

For reasons of consistency, the symbols and general flow used in these next sections follow the wireless communications textbook by Rappaport [Rap02].

4.4.2 Free Space Loss

The free-space propagation model is used to predict received signal strength when the Tx and Rx have a clear, unobstructed LOS path between them. Examples of such systems are satellite communications at higher elevation angles and communications between microwave towers. Free-space communication follows the Friis equation where the received power falls off as the square of the Tx-Rx separation as:

$$P_r(d) = \left[P_t G_t G_r \lambda^2\right]\left[(4\pi)^2 d^2\right]$$

(4.4.1)

Here, if the gain $G_t = 1$, the antenna is regarded as an isotropic radiator. This implies that the received power decays with distance at a rate of 6 dB when the range is doubled, or 20 dB/decade. The product $P_t G_t$ defines the effective isotropic power (EIRP).

In practice, effective radiated power (ERP) is used instead of EIRP; this is the radiated power from a dipole antenna having a numeric gain of 1.64 or 2.15 dB above an isotropic gain. The Friis formula is only valid for received signal P_r in the far field, or Fraunhofer, region beyond the far-field distance and is related to the transmit antenna aperture and carrier wavelength. The Fraunhofer distance d_f is given by $d_f = 2D^2/\lambda$. The received power in free space at a distance d ($d > d_0$) is $P_r(d) = P_r(d_0)(d_0/d)^2$. Here, d_0 is the reference distance greater than d_f and $d \geq d_0 \geq d_f$. The free-space loss equation can also be related to the electric field E at a distance d [Gri87, Kra50]. The equation is given by

$$P_r(d) = P_d A_e = \frac{|E|^2}{\eta} = \frac{|E|^2}{120\pi} A_e = \frac{P_t G_t G_r \lambda^2}{(4\pi)^2 d^2} Watts$$

(4.4.2)

Here, η is the intrinsic impedance of free space and has a value of 120π or 377 ohms

Note that the maximum permissible line-of-sight distance between transmitter and receiver in a mobile system is a function of the heights of the transmit and receive antennas and is limited due to the curvature of the earth. Thus, the free-space path-loss (PL) model cannot be applied beyond this distance.

4.4.2.1 Adjustment of Path Loss Due to Reflection

As stated in section 4.4.1, the free-space path-loss equation applies only under very restricted conditions. In practice, there are usually obstructions in or near the propagation path, or surfaces from which a radio wave can be reflected. An interesting practical case in mobile communications is propagation between two elevated antennas of height h_t and h_r separated by a few tens of kilometers. In this situation, it is permissible to neglect the earth's curvature and assume the surface to be smooth and flat, as shown in Figure 4-1.

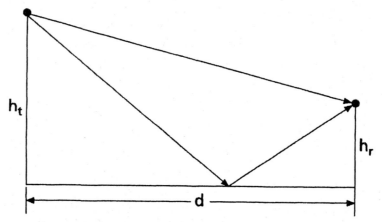

Figure 4-1: Propagation Between Two Elevated Antennas

If $d \gg h_t, h_r$ the two-ray ground bounce model can be expressed as

$$\frac{P_r}{P_t} = G_t G_r \left(\frac{h_t h_r}{d^2} \right)^2$$

(4.4.3)

The above equation in dB is

$$
\begin{aligned}
PL(dB) &= 10 \log_{10}(P_t / P_r) \\
&= -10 \log_{10}(G_t) - 10 \log_{10}(G_r) - 20 \log_{10}(h_t) - 20 \log_{10}(h_r) + 40 \log_{10}(d)
\end{aligned}
$$

(4.4.4)

When applied to a cellular system, this equation suggests the fourth-power relationship of the path loss with the distance from the cell-site antenna, and a 6-dB per octave variation because of antenna heights.

4.4.2.2 Adjustment of Path Loss Due to Diffraction

Sharp edges obstructing the radio path between transmitter and receiver can cause the electromagnetic waves to bend. The resulting loss of radio energy is known as diffraction loss. This loss can be computed by evaluating the Fresnel-Kirchoff diffraction parameter υ, which depends on the height of the ridge above or below the line joining the transmitter and receiver antennas. The resulting loss in dB due to diffraction is computed by:

$$G_d = 20 \log_{10} |F(\upsilon)|$$

(4.4.5)

In practice, graphical or numerical techniques are used to compute diffraction loss. An approximate solution of equation 4.4.5 for different values of υ has been provided by Lee [Lee85]. Bullington [Bul47] provides a method for replacing multiple edges by a single equivalent edge. Various approaches to estimate the diffraction loss for multiple ridges have been proposed by Millington [Mil62] and others [Dey66, Eps53].

As mentioned above, the height of the ridge above or below the line joining the transmitter and receiver antennas is important. The radio waves occupy the volume in space known as the Fresnel zone [Bul77], with the height of the Fresnel zone measured in terms of wavelength λ. In practice, no additional loss (other than free-space loss) is encountered if the ridge is below the direct path by 1λ or more. The loss will increase as the clearance is reduced below λ and the waves will bend (diffraction loss) if the obstruction completely covers the direct path.

4.4.2.3 Adjustment of Path Loss Due to Scattering

Scattering occurs when the medium of transmission has many objects with dimensions smaller than the wavelength. Typically such objects in mobile surroundings include foliage, street signs, and lamp posts. To predict the propagation loss over an arbitrary surface of the earth, the roughness of the ground must be taken into account. The criterion generally accepted for defining the surface roughness is the Rayleigh criterion. This states that if the path-length difference $(r_2 - r_1)$ between wavefronts A and B (on two paths) does not exceed one-quarter of a wavelength then the surface is regarded as electrically smooth [Meh94]. Under this constraint the electrical phase difference from Figure 4-2 is given by:

$$\Delta\psi = (2\pi/\lambda)(2H\sin\theta) \cong \lambda/8\theta \qquad (4.4.6)$$

This equation states that for a small grazing angle, ε, the physical size of the irregularity can be relatively large without destroying the electrical smoothness of the reflecting surface. On the other hand, for the same value of ε at a higher frequency, the irregularity must be kept small. As an example, at 900 MHz and for a grazing angle ε of 1 deg, the computed value for the Rayleigh height is 2.38 m. Unless the irregularity height exceeds 2.38 m, then, the surface will be regarded as smooth. A smoother surface generates higher reflected energy, which lowers the loss.

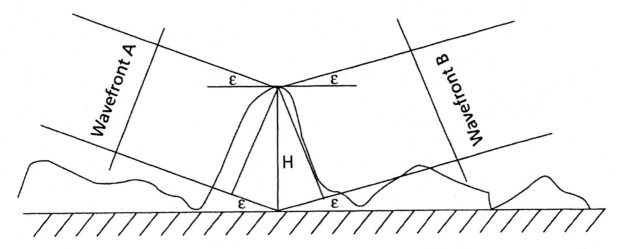

Figure 4-2. Electrical Phase Difference Between Two Wavefronts

4.4.3 Statistical Fading Models

Useful propagation models are mostly based on combining analytical and empirical methods. The empirical approach fits curves derived from measured data. In general, these curves are very useful because they take into account all known and unknown propagation factors through measured data. However, on frequencies. In most cases these curves are re-verified before being used to design the system. Over time, some classical propagation models have emerged that are now used to predict large-scale coverage for mobile communication system designs. Two such models are discussed next.

4.4.3.1 Log-distance Path-loss Model

Models based on both analysis and experimental data reveal that the propagation loss increases as some power of distance. The exponent n varies from place to place and assumes different values in different environments. The range of n is from 2 for free-space propagation to about 6 in urban areas with tall buildings. Thus, choosing the right exponent, n, is necessary for a realistic loss estimate. The average large-scale path loss for a Tx-Rx separation distance, d, with exponent n is given by

$$\overline{PL}(d) \propto \left(\frac{d}{d_0}\right)^n \tag{4.4.7}$$

or

$$\overline{PL}(d)[dB] = \overline{PL}(d_0) + 10n\log_{10}\left(\frac{d}{d_0}\right) \tag{4.4.8}$$

The bars in these equations denote the ensemble average of all possible path-loss values at distance d. The parameter d_0 is the reference distance at the far field of the antenna. The dB plot of this equation is a straight line on log-log paper where the slope is $10n$, or the loss increases by 10 dB with every decade increase in distance. The recommended value of d_0 is 1 km for a large-scale propagation model and 100 m for cellular and microcellular systems [Lee85].

4.4.3.2 Log-normal Shadowing

The above model can be modified to account for clutter around the area of interest. The choice of n and the power variance factor are added to the equation to match the measured data. Measurements have shown that at any value of d, the path loss $PL(d)$ at a particular location is random and distributed log-normally (normal in dB) about the mean distance-dependent value [Cox84, Ber87]. The modified form of the equation is

$$PL(d)[dB] = \overline{PL}(d) + X_\sigma = \overline{PL}(d_0) + 10n\log_{10}\left(\frac{d}{d_0}\right) + X_\sigma \tag{4.4.9}$$

Here, X_σ is the zero mean Gaussian process having a standard deviation of σ, in dB. The equation is valid over a large number of measurement locations having the same Tx-Rx separation but with different clutter factors.

A linear regression model has been applied to find the value of n and σ from measured data in six cities in Germany, an approach that minimizes the error between the predicted loss curve and the measured data [Sei91].

4.4.3.3 Determining the Coverage Area

One can anticipate that the 100% coverage over a designed area may not be possible due to the shadowing effect of the surroundings. Jakes suggested finding the percentage of useful area, $U(\gamma)$, within the circle of radius R where the received signal power will exceed the desired threshold [Jak74]. At a distance $d = r$ at the boundary where the signal will exceed the desired threshold, γ, with certain probability is given by

$$U(\gamma) = \frac{1}{\pi R^2} \int \Pr[\Pr(r) > \gamma] dA = \frac{1}{\pi R^2} \int_0^{2\pi}\int_0^R \Pr[\Pr(r) > \gamma] r\, dr\, d\theta \qquad (4.4.10)$$

Jakes has used the approach to arrive at a family of curves relating a fraction of the total area to signals above the threshold, $U(\gamma)$, as a function of the probability of the signal being above the threshold on the cell boundary.

4.4.4 Outdoor Propagation Models

Radio transmission outdoors depends upon a precise estimate of the terrain undulation, building heights, and the surrounding greenery, such as plants, trees, etc. In urban areas, the average height of the buildings plays a critical role in path-loss determination, while in rural areas shadowing, scattering and absorption by trees and other vegetation can play a vital role, especially at higher frequencies. A large number of models are available to predict path loss over irregular terrain. While all these models aim to predict signal strength at a particular receiving point or in a specific local area (called a sector) the methods vary widely in their approach, complexity, and accuracy. The prediction models differ in their applicability over different terrain and environmental conditions. Some are generally applicable while others are restricted to specific situations. However, no one model stands out as ideally suited to all environments. Thus, the model must be chosen carefully. Most models aim to predict the median path loss, i.e., the loss not exceeded at half the locations and/or for half the time. Knowledge of the signal statistics then allows the variability of the signal to be estimated. This makes it possible to determine the percent of the specified area where the received signal strength is adequate. From the large number of models we will discuss the models by Egli, Okumura, Hata, and COST-231.

4.4.4.1 The Egli Model

Egli's model for median path loss (path loss less than the predicted value at half the locations and half the time) is

$$L_{50} = G_t G_r \left[\frac{h_t h_r}{d^2} \right]^2 \beta \qquad (4.4.11)$$

The factor β accounts for excess path loss and is given by $\beta = (40/f_{MHZ})^2$. Egli's interpretation of β at 40 MHz was to represent median path loss irrespective of the irregularity of the terrain. He further related the standard deviation of β to that of the terrain undulations by assuming the terrain height to be log-normally distributed about its median value [Egl57, Reu74]. Hence, he depicted these results in a set of curves that provide the value of β in dB referenced to the median value versus frequency in MHz.

4.4.4.2 The Okumura Model

One of the most widely used models for predicting path loss in urban areas is Okumura's. The model applies in the frequency range of 150 MHz to 1920 MHz (and extrapolates up to 3000 MHz). The transmitter-to-receiver separation can be from 1 to 100 km and the base antenna height can range from 30 to 1000 m. The median attenuation equation is represented as

$$L_{50}(dB) = L_F + A_{mu}(f,d) - G(h_{te}) - G(h_{re}) - G_{AREA}$$ (4.4.12)

where L_{50} is the 50th percentile (i.e., median) value of propagation path loss, L_F is the free space loss, A_{mu} is the median attenuation relative to free space, and $G(h_{te})$ and $G(h_{re})$ are the base-station and the mobile-antenna gain factors. G_{AREA} is the gain due to the environment. The gain factors of the base and mobile antennas are strictly due to their heights and exclude the effect of their antenna gain patterns. $A_{mu}(f,d)$ is represented by a set of parametric curves where the antenna heights h_t and h_r have fixed values of 200 m and 3 m, respectively. The following equations provide corrections to antenna gains if the base and mobile antenna heights are different from 200 m and 3 m.

$$G(h_{te}) = 20\log_{10}\left(\frac{h_{te}}{200}\right) \quad 30\,m < h_{te} < 100\,m$$

$$G(h_{re}) = \begin{cases} 10\log_{10}\left(\frac{h_{re}}{3}\right) & h_{re} < 3m \\ 20\log_{10}\left(\frac{h_{re}}{3}\right) & 3m < h_{re} < 10\,m \end{cases}$$ (4.4.13)

G_{AREA} corrections are divided into three areas: open, quasi-open and suburban. Correction factor in dB are read as a function of the operating frequency. Other correction factors applicable to the Okumura model involve the terrain undulation height, Δh, the average slope of the terrain, the isolated ridge height, and the mixed land-sea path parameter [Oku68].

The wholly empirical nature of the Okumura model means that the parameters used are limited to specific ranges determined by the measured data. If, in attempting to make a prediction, the user finds that one or more parameters fall outside the specified range then there is no alternative but to extrapolate the appropriate curve, which may give unrealistic results. Care must be exercised.

4.4.4.3 The Hata Model

To make the Okumura method easy to apply, Hata established empirically-based mathematical relationships to describe the graphical information given by Okumura. Hata's equation applies directly to urban-area propagation as a standard formula and supplies corrections to other situations. The formula for urban surroundings is

$$L_{50,urban} = 69.55 + 26.16\log_{10}(f_c) - 13.82\log_{10}(f_c)$$
$$- 13.82\log_{10}(h_{te}) - a(h_{re}) + 44.9 - 6.55\log_{10}(h_{te}) + 10\log_{10}(d)$$ (4.4.14)

Here, the frequency range is from 150 to 1500 MHz, h_{te} is the effective transmitter (base station) antenna height (in meters) ranging from 30 to 200 m, and h_{re} is the effective receiver (mobile) antenna height (in meters) ranging from 1 to 10 m. The factor d is the Tx-Rx separation distance (in km), and $a(h_{re})$ is the correction factor for the effective mobile antenna height, which is a function of the size of the coverage area. Hata provides mobile antenna correction factors in dB for small and medium cities [Hat90] as

$$a(h_{re}) = (1.11 \log_{10} f_c - 0.7) - (1.56 \log_{10} f_c - 0.8) \tag{4.4.15}$$

and for large cities as

$$a(h_{re}) = \begin{cases} 8.29 (\log_{10}(1.54 h_{re}))^2 - 1.1 & f_c \le 300 MHz \\ 3.2 (\log_{10}(11.75 h_{re}))^2 - 4.98 & f_c > 300 MHz \end{cases} \tag{4.4.16}$$

Hata also modified this last equation for suburban and open rural areas. These equations are

$$L_{50}(dB) = \begin{cases} L_{50,urban} - 2[\log_{10}(f_c/28)]^2 - 5.4 & Surburban \\ L_{50,urban} - 4.78[\log_{10}(f_c)]^2 + 18.33 \log_{10}(f_c) - 40.94 & Open \& Rural \end{cases} \tag{4.4.17}$$

Although Hata's model does not have any of the path-specific corrections of Okumura's model, these expressions have significant practical value. Results using the Hata model compare very closely with the original Okumura model, as long as d exceeds 1 km. Thus, the model provides good results for large cells but is not applicable to the smaller PCS-type cells.

4.4.4.4 The COST-231 Model

The high end of the frequency range suitable for the Hata model is 1500 MHz. This limits its application to the PCS range. Under the European COST-231 program, the Okumura model results were analyzed at the upper end of its frequency range. Modifications were then made to the program so it can be applied between 1500 and 2000 MHz. The PL model is

$$L_{50}(Urban) = 46.3 + 33.9 \log_{10}(f_c) - 13.82 \log_{10}(h_{te}) - a(h_{re})$$
$$+ 44.9 - 6.55 \log_{10}(h_{te}) \log_{10} d + C \tag{4.4.18}$$

Here, the values of $a(h_{re})$ are the same as in the Hata model. The value of C is 0 dB for medium-size city and suburban areas and 3 dB for metropolitan areas. The model applies with the following limits:

$$\begin{aligned} 1,500\,MHz &\le f_c \le 2,000\,MHz \\ 30m &\le h_{te} \le 200m \\ 1m &\le h_{re} \le 10m \\ 1Km &\le d \le 20\,Km \end{aligned} \tag{4.4.19}$$

4.4.4.5 Ray Tracing

The availability of high-resolution databases makes it attractive to move to deterministic propagation methods. One such method based on ray theory can be applied for both indoor and outdoor environments. If a number of rays can be traced from a given transmitter to a given receiver, the electrical lengths of various ray paths give the amplitude and phase angle at the receiver, which can be used to estimate the propagation loss. In evaluating each path one must account for direct as well as indirect reflection paths from obstructions. By including the antenna characteristics at both ends one can estimate the complete channel characterization of the propagation. The method can be used for both outdoor and indoor modeling. In outdoor applications, sophisticated processing techniques can be used to convert aerial or satellite photographs into 3D databases [Rus93, Sch92, Wag94]; in indoor environments, architectural drawings and other layout information can serve the same purpose [Val93, Sei94, Kre94, Mor00].

Two basic methods have appeared in the literature, the ray launching or "brute-force," method and ray tracing [Sei92, McK91]. In ray launching, a software program checks for a LOS path between a specified transmitter and receiver. Next, it launches and traces a ray away from the transmitter in the specified direction and detects whether it intersects an obstruction specified in the database. If it does not, the process stops and a new source ray is launched in a different direction. If an intersection is found, the program determines whether the reflected ray from the intersection point has an unobstructed path to the receiver; the reflected and transmitted rays are then traced to the receiver or to another obstruction. This process continues for each ray until the ray reaches the receiver, until a specified number of obstructions are exceeded, until no further intersections occur, or until the energy level falls below a specified threshold. In this method, all possible launch angles in 3D space, each having a small separation with its neighboring path, must be considered. A tradeoff between coverage and computation time limits the angular separation to about 1^0 [Sch94].

Unlike the ray-launch method, the image-based approach considers all obstructions as potential reflectors and calculates their effect. This is strictly an analytical approach where paths are not duplicated, nor left out. In simple environments the computation time is much less because only paths are considered that actually exist between the transmitter and the receiver. To make the process more meaningful, a wall sequence diagram is created. Figure 4-3 illustrates the case of four walls within which both the transmitter and receiver are located. Here, wall 1 is the first reflector (one single reflection, three secondary reflections, and nine tertiary reflections). As many as 13 reflections are possible. Similarly, the reflection can start with wall 2, 3, or 4. Thus, a complete analysis must consider 52 paths.

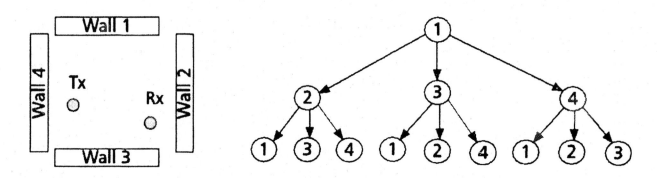

Figure 4-3: Enumeration of Reflection Paths for a Four-Wall Test Case

4.4.4.6 Outdoor Terrain

Propagation curves must account for environmental clutter and terrain irregularities correctly so that the predicted field strength comes close to the measured value of the propagation loss. There are infinite varieties of terrain features, but for reference analysis "quasi-smooth" terrain is assumed. Whether in quasi-smooth or irregular terrain, buildings and trees near the vehicular antenna affect the received field strength differently. If we classify obstacles in areas according to buildings or trees, it will make the prediction of field strength in a given area very difficult. Thus, rather than classifying them into many minute groups, the clutter is generally classed according to the degree of congestion in a given area, i.e., an urban area that has large buildings, further divided into dense urban, moderate urban, and geographically limited urban. The terrain classification is shown in Table 4-1.

The terrain undulation height Δh for rolling hills is defined by Okumura as the difference between 10% and 90% of the terrain undulation height within a distance of 10 km in front of a mobile antenna. A similar characterization by the International Consultative Committee on Radio (CCIR) defines the parameter Δh as the difference between the elevation exceeded by 90% of the terrain and the elevation exceeded by 10% of the terrain some 10 to 50 km in front of the transmitter. Both these definitions are shown in Figure 4-4 and Figure 4-5. Okumura also defines the average slope as θ (positive or negative) measured over 5–10 km. In his model, Okumura applies a rolling-hill correction factor, K_{ter}, in dB for different terrain undulation heights.

Category	Characteristics
Open area	Very few structures, such as tall buildings and trees, are in the propagation path. The area is mostly farms, open fields, rice fields, etc.
Suburban area	Residential, with small one or two-story houses, with some obstacles near the mobile radio, but not very congested
Urban	Heavily built-up areas with tall buildings and high-rise apartments

Table 4-1: Characteristics of Different Terrain

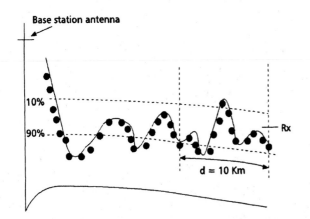

Figure 4-4: Definition of Terrain Undulation Height

Figure 4-5: CCIR Definition of Parameter Δh for Rolling Hilly Terrain

4.4.4.7 Indoor Structure: Wall

The modeling of radio-wave penetration into the outer walls of a building differs from the more familiar cellular case in many respects. The modeling here is based on transmitters mounted outside on a high-rise building while the mobile unit is located inside a different building and at a varying height. In a rural environment this may result in a LOS path to the upper floors of many buildings, whereas this is a relatively rare occurrence in buildings located in urban and suburban surroundings.

In typical cellular systems, base stations for macrocells are located on the roofs of tall buildings, and may be 100 m or more above the local terrain; the mobile can be a few kilometers away. Since the cellular model setup is different from the model here (inside the building), it is difficult to use the results from the cellular design directly.

Penetration loss experiments have shown that the signal in small areas within buildings is approximately Rayleigh distributed, with the scatter of the medians having a log-normal distribution. In other words, the signal statistics within a building can be modeled as superimposed small-scale (Rayleigh) and large-scale (log-normal) processes.

Measured penetration loss has been found to be a function of frequency as well as height within the building, as shown in Table 4-2. The antenna pattern in the vertical plane also plays a role in how much signal penetrates a building. Measured values of penetration loss at 441.0, 869.5, and 1400 MHz in a six-story building in Liverpool, UK [Tur87] shows that the penetration loss is high at ground level but decreases for higher floors. Also, the penetration loss decreases as frequency increases. Measurements by Turkmani showed penetration values of 14.2, 13.4, and 14.2 dB at 900, 1800, and 2300 MHz, respectively [Tur92], consistent with measurements at Liverpool.

Floor Level	Penetration Loss (dB)*		
	441.0 (MHz)	896.5 (MHz)	1400.0 (MHz)
Ground	16.37	11.61	7.56
1	8.11	8.05	4.85
2	12.76	12.5	7.98
3	13.76	11.18	9.11
4	11.09	8.95	6.04
5	5.42	5.98	3.31
6	4.2	2.53	2.54

* Measured values are with respect to signal measured outside the building

Table 4-2: Penetration Loss Versus a Building's Floor Level

Received signals at different floors and at different points on the same floor can be measured and the penetration loss estimated. This survey is necessary from a performance perspective and is somewhat similar to that discussed under "RF Site Surveys" in section 4.3.10.

4.4.4.8 Partition Losses between Floors

It has been reported that losses between floors are influenced by the construction materials in the outside walls, the number and size of windows, and the type of glass. The external surroundings also must be considered since there is evidence [Dev90] that energy can propagate outwards from a building, be reflected and scattered from adjacent buildings, and re-enter the building at a higher and/or lower level depending on the location of the antenna. Experiments have also shown that attenuation between adjacent floors is greater than the incremental attenuation caused by each additional floor but that after five or six floors there is little further attenuation. The floor attenuation factor (FAF) and the standard deviation σ (dB) for three buildings have been tabulated by Rappaport [Rap02] and Siedel [Sei92]. Values for Pacific Bell's building in San Francisco range from 13.2 dB of loss on the first floor to 27.1 dB on the fifth floor.

The standard deviation values are 9.2, 8.0, 5.6, 6.8, and 6.3 dB for the first through fifth floors. The measurement process is similar to the site surveys discussed in section 4.3.10.

4.4.4.9 Partition Loss on Same Floor

Building partitions take a different form based on whether the building holds residences or offices. Residential buildings use wood-frame partitions with sheetrock and generally have wooden floors. On the other hand, office buildings use either moveable office partitions (soft) or permanent (hard) partitions as part of the building structure. Materials used vary so widely that a general model for partitions does not exist. However, values of the loss caused by different types of partition materials have been tabulated by various authors [Rap91, Cox83, Vio88]. For example, the loss with frequency for materials such as 3/8-in.-thick sheetrock (2 sheets) is 2 dB at 9.6 GHz. Similarly, the measured loss for ¾-in.- thick dry plywood (1 sheet) is 1 db at 9.6 GHz [Rap02]. The measurement process was similar to the one in the site- survey discussion in section 4.3.10.

4.4.4.10 Path Loss Within Building

A path-loss model in dB based on the floor attenuation factor (FAF), partition attenuation factor (PAF) for some specific obstruction, and the exponent value for the "same floor" (n_{SF}) have been provided by Seidel [Sei92] as

$$\overline{PL}(d) = \overline{PL}(d_0) + 10 n_{SF} \log_{10}\left(\frac{d}{d_0}\right) + \text{FAF} + \sum \text{PAF} \qquad (4.4.20)$$

A modified form of this equation that includes the loss term due to a multistory structure has been suggested by Devasirvatham, et al. [Dev90] as

$$\overline{PL}(d) = \overline{PL}(d_0) + 20 \log_{10}\left(\frac{d}{d_0}\right) + \text{FAF} + \sum \text{PAF} + \alpha d \qquad (4.4.21)$$

Here, α is the attenuation constant for the signal channel in dB per meter. Measured values of α for a two- and four-story building have also been tabulated by Devasirvatham.

4.4.4.11 Non-Line-Of-Sight Loss (NLOSL) for Ultra-Wideband Channel

A report by Buehrer et al. documents the results of propagation and delay studies conducted for ultra-wideband systems (UWBs) by various commercial organizations, including a study conducted by Virginia Tech [Bue04].

Path Loss Study

Traditionally, path-loss analysis is based on the Friis transmission formula, which depends upon the operating frequency and path length. The Buehrer study attributes the path loss to the antenna and is independent of operating frequency. The study measures the path loss and its variance with respect to free-space loss at a distance of 1 m from the transmitter. Equations describing the average path loss and the received signal power on which the experimental results are based are

$$\overline{PL}_{dB} = 10\log10\left(\frac{\overline{P_r(d_0)}}{P_r(d)}\right) = -10n\log10\left(\frac{d_0}{d}\right) \tag{4.4.22}$$

$$\overline{P}_r(d)_{dB} = \overline{P}_r(d_0)_{dB} - 10n\log10\left(\frac{d}{d_0}\right) + X_\sigma \tag{4.4.23}$$

Antennas used in the study were a wideband biconical and a TEM horn. Measurements were made at distances of 9 m and 4 m. Results for the mean-path-loss exponent and the standard deviation of the received signal power for both LOS and NLOS cases conducted by various organizations are shown in Table 4-3. In all cases, values for NLOS are greater than LOS values.

Organization	n (mean)	N (Std. Dev.)	σ (dB; mean)	σ (dB; Std. Dev.)
Virginia Tech	1.3–1.4 (LOS) 2.3–2.4 (NLOS)		2.5–3 (LOS) 2.6–5.6(NLOS)	
AT&T	1.7/3.5 (LOS/NLOS)	0.3/0.97	1.6/2.7	0.5/0.98
France Telecom	1.5/2.5 (LOS/NLOS)			
Intel	1.7/4.1(LOS/NLOS)		1.5/3.6(LOS/NLOS)	

Table 4-3: Values Measured in Path Loss Studies

Test results of the mean-excess delay and the RMS delay-spread values $\overline{\tau}$ and σ, are shown in Table 4-4 for both bicone and TEM antennas when threshold values are 15 and 20 dB, respectively. In most cases, values an order of magnitude higher are observed in NLOS compared to LOS values.

	Bicone				TEM			
	15		20		15		20	
	NLOS	LOS	NLOS	LOS	NLOS	LOS	NLOS	LOS
Mean excess delay	1.60E-08	5.19E-09	2.01E-08	1.05E-08	2.36E-09	5.52E-10	5.59E-09	1.22E-09
Maximum excess delay	6.57E-08	2.84E-08	7.86E-08	5.68E-08	1.61E-08	2.65E-09	4.31E-08	1.24E-08
RMS delay spread	1.37E-08	5.41E-09	1.62E-08	8.5E-09	3.27E-09	7.53E-10	7.09E-09	1.70E-09

Table 4-4: Low Altitude Small-Scale Results for Bicone and TEM Antennas

4.4.4.12 Indoor and Outdoor Coverage Calculations

A set of examples illustrating the computational aspect of path loss calculation in mobile surroundings are given below. The examples cover (1) free-space loss; (2) loss due to reflection from the ground; (3) propagation loss inside a building; (4) application of the COST-231 model for path loss calculation; and (5) diffraction loss due to a knife edge.

(1) Compute the free-space loss with the following parameters: P_t = 20 Watt, G_t = G_r = 0.0 dB, f_c = 900 MHz. Find P_r (received power) at a distance of 1.5 km.

$$P_r(d) = \frac{P_t G_t G_r \lambda^2}{(4\pi)^2 d^2} = \frac{20 \times 1 \times 1 \times (1/3)^2}{(4\pi)^2 \times (1.5 \times 1000)^2}$$

$$= \frac{2.22}{355.305 \times 10^6} = 0.006248 \times 10^3 \; mWatt = -52.04 \; dBm \qquad (4.4.24)$$

(2) Two-path model: For the following conditions, test whether the two-ray model can be applied or not: h_t = 40 m, h_r = 3 m, and d = 300 m, and 500 m. When $d > 10(h_t + h_r)$, we can say $d \gg h_t + h_r$. Applying the distance criterion, it fails for d = 300. For d = 500 the criterion is satisfied and the two-ray model applies.

(3) Measurements inside a building follow a log-normal distribution about the mean. Assume the exponent to be -3.5. If a signal of 1 mW was received at d_0 = 3 m from the transmitter and at a distance of 30 m, and 20% of the measurements were stronger than -25 dBm, define the standard deviation σ, for the path loss model at d = 15 m.

$$\overline{Pr(d)} = -35 log10(d/d0) + Pr(d_0) = -35 log_{10}(30/3) + 0 = -35 \; dBm$$

$$For \; \gamma = -35 \; dBm, \; Pr[Pr(d) > \gamma] = Q\left[\frac{\gamma - \overline{Pr(d)}}{\sigma}\right] = Q\left[\frac{-25 + 35}{\sigma}\right] \qquad (4.4.25)$$

$$= Q\left[\frac{10}{\sigma}\right] = 0.2; \; or \; \frac{10}{\sigma} = 0.8416; \quad \sigma = 11.9 \; dB$$

(4) Use the COST-231 model for a large city to find the received power at 2 km and 10 km when the received power at a distance of 1 km = 1 μW. Assume f = 1800 MHz, h_t = 40 m, h_r = 3 m, G_t = G_r = 0 dB.

$$L_{50}(Urban \; at \; d = 2km) = 46.3 + 33.9 log_{10}(1800) - 13.82 log_{10}(40)$$
$$- 3.2(log_{10} 11.75 \times 3)2 + 4.98 + (44.9 - 6.55 log_{10}(40)) log_{10}(2) + 3 \qquad (4.4.26)$$
$$= 145.19 \; dB$$

Similarly,

$$L_{50}(Urban \; at \; d = 1km) = 134.83 \qquad (4.4.27)$$

Thus, the power received at 2 km is

$$P_r(d = 2km) = P_r(d = 1km) + (L_{50}(d = 2) - L_{50}(d = 1)) = -30(dBm) - (145.19 - 134.83) = -40.36 \; dB \quad (4.4.28)$$

Similar to the above calculation for a distance of 10 km, $L_{50}(d = 10 \; km) = 169.24$. Thus, the power received at 10 km is

$$P_r(d = 1km) + (L_{50}(d = 10) - L_{50}(d = 1)) = -30(169.24 - 134.83) = -64.41 \; dBm \qquad (4.4.29)$$

(5) Compute the diffraction loss due to a knife-edge ridge and the value of the power received for the configuration shown in Figure 4-6. Compute also the free-space loss as if the obstruction does not exist. The following parameters can be assumed: P_t = 10 watt, G_t = 10 dB, G_r = 0 dB, and L = 1 dB at f = 900MHz.

$$P_r(d) = \frac{P_t G_t G_r \cdot \lambda^2}{(4\pi)^2 d^2} = \frac{10 \times 1 \times 1 \times (1/3)^2}{(4\pi)^2 \times (6 \times 1000)^2} = \frac{1.11}{5684.9 \times 10^6} = 1.954 \times 10^{-10} \; Watt = -67.1 \; dBm \qquad (4.4.30)$$

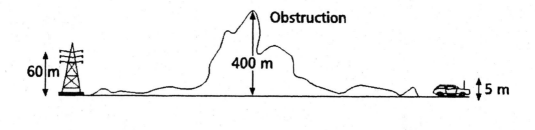

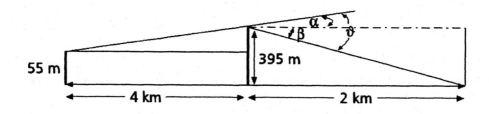

Figure 4-6: Diffraction Loss Computation Due to a 400-m-High Ridge

$$\tan(\beta) = \frac{395}{2000}, \; thus, \beta = 0.195 \, radians; \tan(\alpha) = \frac{340}{4000}, thus, \; \alpha = 0.085 \, radians; \; or \; \theta = \alpha + \beta = 0.275 \, radian$$

$$\upsilon = \alpha \sqrt{\frac{2 \times d_1 \times d_2}{\lambda \times (d_1 + d_2)}} = 0.275 \sqrt{\frac{2 \times 2000 \times 4000}{0.333 \times (6000)}} = 0.275 \times 89.44 = 24.6. \qquad (4.4.31)$$

Using Lee's approximation for $\upsilon > 2.4$, *Diffraction loss* $= G_d = 20 log \left(\frac{0.225}{24.6} \right) = -40.8 \, dB$

Received power $= -67.1 - 40.8 = -107.8 \, dBm$

4.4.4.13 Software Modeling Tools

A large number of simulation models are being marketed for cellular and other mobile surroundings. Following are the salient characteristics of two types of these models: one for designing a wireless outside system in an urban setting and the other for a system inside a typical building.

Modeling an Open Area in an Urban Setting

These models consider the effects of buildings and terrain on the propagation of electromagnetic waves through the wireless system. They predict how the location of the transmitters and receivers in the urban area affect signal strength. The models account for the physical characteristics of uneven terrain and building features, perform electromagnetic calculations, and then evaluate the signal-propagation characteristics. Editing tools can be used to construct a virtual building and terrain environment, or the virtual model can be imported with standard software. Transmitter and receiver locations can be specified using site-defining tools or by importing them from an external data file. Urban and terrain features of large areas may be specified. Separate calculations for portions of the overall area may be specified by defining study areas. These models have been used as validation exercises for large cities, including Ottawa, Canada and Munich, Germany. Results of the Ottawa validation have been reported by J. H. Whitteker [Whi88].

Modeling inside a Typical Building

In producing a 3D simulation, the ray-tracing engines of these software tools generally consider an implementation based on physical and geometric optics and diffraction physics. A wide variety of antenna types can either be chosen from a list or the user can choose an antenna from a pattern file. The ray-tracing process predicts the local mean power received at a given x, y, and z point. For each point the vector sum of the multipath power is computed. Calculations also include the effects of frequency, polarization, material properties, and varying antenna patterns. If desired, the model can include the effects of ray diffraction. Key input parameters for the simulation software include the building facet and edge files; user-defined building materials; antenna patterns, placement, strength, frequency, polarization, and the observation grid

4.4.5 The JRC (Durkin's) Model (a Case Study)

A classical approach relying on terrain data for predicting field strength contours over irregular ground has been adopted by the Joint Radio Committee (JRC) in the U.K .Terrain data is readily available in most of the developed countries, as an example in the U.S. from Geological Survey of America. A computer-based approach forms the basis of almost all design-modeling tools. Though the approach only predicts large-scale phenomena (i.e., path loss), it provides an interesting perspective for estimating the loss caused by obstacles and irregular terrain. The heart of the approach is the elevation information in the data base stored in row and column matrix form, which, in the original version, provided height reference points at 0.5 km interval, as shown in Fig. 4-7.

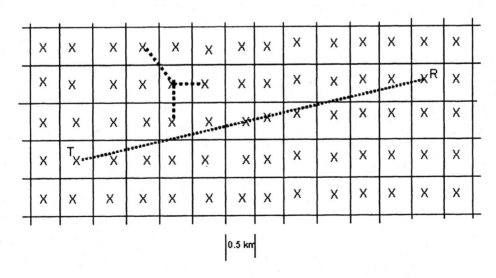

Figure 4-7: Stored Values of Terrain Heights at Different Locations in a Database.

The computer program uses this topographic data to reconstruct the ground-path profile between the transmitter and the chosen receiver, as shown in Figure 4-7, using the row, column, and interpolated terrain heights. Next, the computer tests for the existence of a line-of-sight path and the adequacy of Fresnel zone clearance over that path. If both of these conditions are satisfied, the larger of the free-space and plane-earth losses is taken as a path-loss value, i.e., in these circumstances

$$L(dB) = \max\ (L_F,\ L_p) .\qquad\qquad (4.4.32)$$

If no line-of-sight path exists or if there is an inadequate Fresnel-zone clearance, the computer then estimates the diffraction loss L_D along the path and computes the total loss as

$$L(dB) = \max(L_F\ L_p)\ +\ L_D \qquad\qquad (4.4.33)$$

For estimating L_D, the computer uses the Epstein-Peterson construction for up to three edges [Eps53]. If more than three obstructions exist between transmitter and receiver, an equivalent single virtual knife-edge based on Bullington's approach is constructed [Bul47].

This method is attractive because it models a site-specific propagation computation based on reading data from the digital elevation map. It can produce a signal strength contour that has been reported as good to within a few dB. The disadvantages are that it does not account for losses due to foliage or to buildings and other fabricated structures. The method also does not account for multipath losses other than ground reflection. Thus, these loss factors have to be added externally/manually.

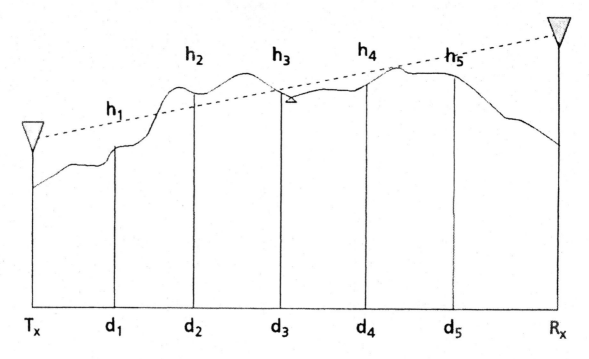

Figure 4-8: Reconstruction of a Terrain Profile

4.4.6 Conclusions

The transmission medium plays a vital role that places a fundamental limit on the performance of mobile communications. Major factors influencing propagation are the effects of irregular terrain and the influence on signals of trees, buildings, and other natural and man-made obstacles. Signal properties inside buildings must also be considered. A brief summary of propagation-measurement methods has been presented here that includes both outside and indoor models. For outdoor applications we have briefly discussed models by Egli and Okumura and variations on the latter by Hata and COST-231.

In addition, a ray-tracing scheme applicable to both indoor and outdoor surroundings was briefly described, and we considered the influence of walls, floors, partitions, etc., on signals inside a building. A case study by JRC (Durkin's model) based on terrain data was outlined. Some numerical examples based on the models have been provided. Since it is impossible to cover every important aspect of propagation in a few pages, a large number of references have been included and the reader is encouraged to study them to enhance understanding.

4.5 RF Engineering

4.5.1 Introduction to Radio Communications Systems

Radio systems, shown in a simplified block diagram in Figure 4-9, are used to transfer information from a source to a remote destination Typically, such a system consists of an information source, which can be audio, video, or computer data, or some combination thereof. A transmitter modulates this information onto a carrier frequency and the result is radiated into the channel.

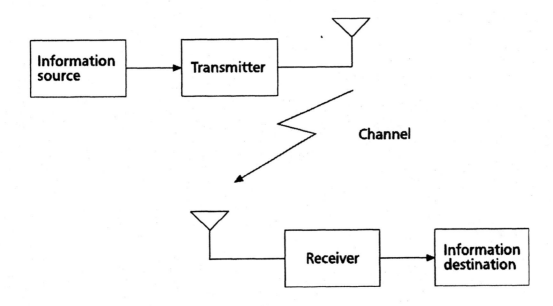

Figure 4-9: A Typical Communications System

The channel is, of course, not perfect. As the signal radiates from the transmitter, its energy density decreases. In addition, the channel, which gains noise from natural and artificial sources, experiences "spreading loss" or a reduction in energy. Also, the frequencies of other transmitters may occupy the channel and add co-channel or adjacent-channel interference.

The receiver recovers the information from the channel. While our perspective in this section is of receivers for VHF/UHF wireless applications, the principles apply to any radio receiver. The ideal receiver has a number of characteristics:

Sensitivity: Radio signals that reach the receiver are extremely weak, so the ideal receiver should be able to detect very small signals.

Low noise: Noise and sensitivity are closely related because noise dictates the weakest detectable signal. At frequencies in the low-VHF range and below, receiver sensitivity tends to be limited by atmospheric noise sources. In the UHF range and above, sensitivity tends to be limited by internally generated amplifier noise.

Selectivity: The ideal receiver should detect only the desired signal and reject all others.

Stability: The receiver should remain tuned to the desired frequency, and its gain, noise, and demodulation characteristics should remain constant despite variations in supply voltage, temperature, vibration, and other environmental conditions.

Low spurious responses: Certain receiver topologies have undesirable responses at specific frequencies. It is possible to predict these frequencies and minimize these responses.

Low phase noise: Many receiver topologies employ one or more internally generated signals. Noise can affect the oscillators that generate these signals, resulting in phase noise. High phase noise can reduce the dynamic range of a receiver, as well as interfere with the detection process for phase-sensitive modulations such as quadrature amplitude modulation (QAM).

Wide dynamic range: It is desirable that the receiver detect weak signals close in frequency to strong ones.

Automatic gain control (AGC): In some receiver systems, it is desirable that the output remain constant for varying inputs. The ideal AGC system maintains the output constant over a wide range of input signal levels and follows temporal input variations without overshoot or delay.

The various issues in receiver design are covered in the literature [Gos86, Roh88, Sab87]. For those oriented toward circuit design, the amateur radio literature has some excellent resources [Hay04, Hay03].

4.5.2 Receiver Topologies

4.5.2.1 The Superheterodyne Receiver

The most widely used receiver topology is the superhetrodyne, or superhet, whose block diagram is shown in Figure 4-10. The incoming signal from the desired channel passes through a pre-selector filter, an RF amplifier, and an image filter, where it is applied to the mixer, which acts as a frequency converter. The signal is combined with the local oscillator (LO) in the mixer. The output of the mixer is nominally the sum and difference of the signal and LO frequencies. The intermediate-frequency (IF) filter selects either the sum or the difference frequency as the intermediate frequency. The IF amplifier provides additional gain at the IF and its output is applied to the detector/demodulator, which extracts the original information.

This design has numerous advantages. Gain can be distributed between the RF or signal frequency and the IF to produce a very-high-gain yet stable receiver. Moreover, the receiver can be easily tuned to a new channel by changing the frequency of the LO. The IF filter generally has a fixed center frequency so it can be optimized for bandwidth and/or delay characteristics, and it does not have to be retuned when the receiver is tuned to a new channel.

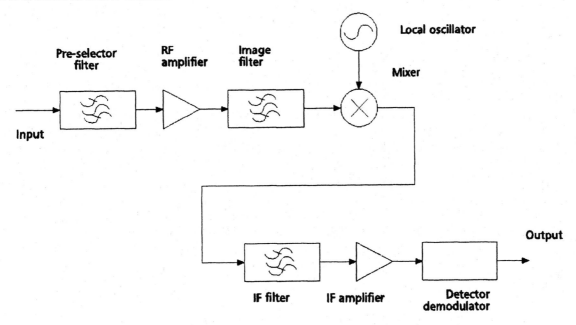

Figure 4-10: Block Diagram of a Superhetrodyne Receiver

4.5.2.2 Receiver Sensitivity

At UHF, internally generated noise limits the weakest signal that can be detected. Noise generated by the RF amplifier is amplified by all the stages in the receiver so this noise is what determines the receiver sensitivity. Low-noise, high-gain amplifiers for the RF amplifier produce the highest sensitivity.

Receiver sensitivity is measured in a number of ways but is typically expressed as the noise figure (NF) and noise temperature (T). The NF is defined as the ratio of the input signal-to-noise ratio (S/N) to the output signal-to-noise ratio. The output S/N is measured at the input of the demodulator. The NF must be 1 or greater because the intervening stages can only degrade the S/N. The NF is usually defined in units of dB but calculations involving the NF must be done as ratios.

Referenced to the antenna port, there is a signal input power, P_{Sin}, and a noise power, P_{Nin}. Any electronic device not at absolute zero creates noise over a wide bandwidth due to black-body radiation. The input noise power comes from losses in the antenna, atmospheric noise, and sky or ground noise. This noise is usually lumped into an equivalent antenna noise temperature, T_A and the input noise is given by $P_{Nin} = kT_A B$, where B is the system bandwidth and k is Boltzmann's constant ($1.38 \times 10^{-23} \, Joules/Kelvin$).

Many active devices will add noise power far different from what their physical temperature might suggest. They are assigned an equivalent noise temperature, T_e. This temperature equates the device noise with a resistor having the equivalent physical temperature that produces the same amount of noise. The NF is calculated as

$$NF \equiv \frac{(S/N)_{in}}{(S/N)_{out}} = \frac{P_{Sin}/P_{Nin}}{GP_{Sin}/G(P_{Sin}+kT_eB)} = \frac{P_{Sin}/kT_AB}{GP_{Sin}/G(kT_AB+kT_eB)} = 1 + \frac{T_e}{T_A} \qquad (4.5.1)$$

where G is the gain of the receiver. Note that the NF is not determined by the signal power, the gain, or the system bandwidth. For this reason, the NF is usually calculated at the receiver input. The relation between the NF and T_e is given by

$$T_e = 290(NF-1) \qquad (4.5.2)$$

Note that this assumes that T_A = 290 K. For satellite systems, T_A may be 50K since the antenna will be pointed at a "cold" sky while, for terrestrial systems, much of the antenna beam will intersect a "hot" earth and T_A will be more like 290K. Historically, the NF has been tied to terrestrial systems and hot-cold load measurements where one load is at room temperature.

While the NF and noise-temperature analysis for cascaded stages is found in many textbooks, an excellent analysis that handles the NF of lossy devices and the correction necessary to handle additional noise produced by spurious mixer products has been performed by Maas [Maa92]. The image filter and the pre-selector filter affect receiver noise performance differently. The image filter is needed to reduce the effect of RF amplifier noise at the mixer image response. The image response is addressed in the next section.

4.5.2.3 Mixing and the Generation of Spurious Responses

The mixing or frequency-conversion process introduces spurious responses. The main spurious response is called the image. Consider a simple example: a TV broadcast receiver is tuned to 627.25 MHz, which is the video carrier frequency for channel 40. The receiver has an IF of 45 MHz so the LO runs at 582.25 MHz. A difference frequency of 45 MHz is generated by mixing 627.25 MHz and 582.25 MHz. Note that when 537.25 MHz, which is the video carrier for channel 25, is mixed with 582.25 MHz, the difference is also 45 MHz. The receiver responds to channel 25 just as it does to channel 40. Figure 4-11 shows the frequency relationships of the desired signal, the image, and the LO. The function of the pre-selector and image filters in the figure is to pass the desired signal but reject the image.

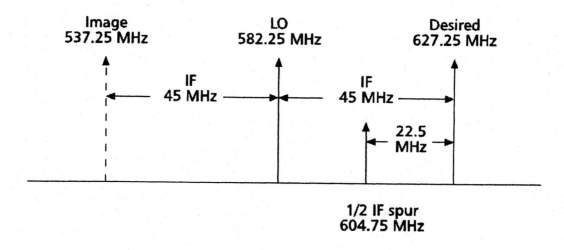

Figure 4-11: Spurious Mixing in a Superhet Receiver

The image is only one of many possible spurious mixing responses. In practice, the mixer is a highly nonlinear device that generates harmonics of the applied LO and signal frequencies. The second harmonic of the LO at 582.25 MHz is 1164.5 MHz. The second harmonic of 604.75 MHz is 1209.5 MHz. If a 604.75-MHz signal reaches the mixer, the sum and the difference of harmonics are created. he difference is the IF of 45 MHz, and the receiver has a spurious response at 604.75 MHz. This response is sometimes called the "half IF spur" because it appears only 22.5 MHz, or one half the IF frequency, from the desired response. The image is twice the IF or, in this case, 90 MHz. It is much easier to design the image and pre-selector filters to remove the image than the half IF spur.

Spurious mixing products are a serious problem in superhet receivers. Mixers generate not only the second harmonic but higher order harmonics as well. Balanced mixers that suppress even-order harmonics are widely used to reduce some of the spurious mixing products. It is also possible for superhet receivers to have more than one mixer, but predicting all the spurious mixing products can be difficult and tedious. Careful selection of IFs, mixers, and filters reduces the impact of spurious mixing. Commercial software is available to aid in predicting spurious responses.

In addition, the LO may leak out through the image filter, the RF amplifier, and pre-selector filter and radiate from the antenna input. This can create interference. An example of this is the LO in TV and FM broadcast receivers that tunes through portions of the aircraft communications band. Thus, operation of these receivers is banned on commercial aircraft.

4.5.2.4 Dynamic Range

At UHF frequencies it is difficult to build a pre-selector filter narrow enough to pass only a single channel of information. In addition, it may be undesirable to build very narrowband filters if the receiver is to be tuned over a range of frequencies. If the range of desired signal frequencies cannot pass through the pre-selector/image filters, then these filters must be retuned when the LO is tuned to the new frequency. This is called tracking and it increases the receiver's complexity.

Since most pre-selector/image filters will pass more than the desired signal, many different signals will be amplified by the RF amplifier and applied to the mixer. All these signals will produce LO sum and difference frequencies. The IF filter will pass only the desired intermediate frequency. However, real mixers will create harmonics of all the applied signals and produce sum and difference frequencies of these harmonics. The result may be a mixing product that falls in the IF filter's bandwidth. The half IF spur shown in Fig. 4-11 is an example.

Similarly, RF amplifiers are not perfectly linear and strong signals can cause them to saturate. This nonlinearity generates harmonics that mix either in the amplifier itself or in the following mixer. While careful design can minimize internally generated mixing products, it is impossible to completely eliminate them with real mixers and amplifiers.

The linearity of amplifiers and mixers is specified in terms of third-order mixing products. Figure 4-12 shows a test setup for measuring these products. The output for an ideal amplifier would be only f_1 and f_2, but nonlinearities in actual amplifiers generate harmonics of f_1 and f_2 *plus* the mixing products of these harmonics. These mixing products are specified in terms of the harmonic number of the signals that generate them. For example, the third-order products are $2(f_1)-f_2$ and $2(f_2)-f_1$. Figure 4-13 shows the output of a non-ideal amplifier. The odd-order products are the most problematic since they fall close to the desired signal, making them difficult to filter out.

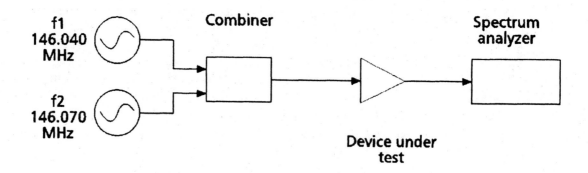

Figure 4-12: Measuring a Third-Order Intercept

Consider a channelized FM mobile-radio system. Channel 1 is 146.01 MHz, channel 2 is 146.04 MHz, channel 3 is 146.07 MHz, and channel 4 is 146.10 MHz. The second harmonic of channel 2 is 292.08 MHz. If this mixes with channel 3, the difference is 146.01 MHz, which is channel 1. Similarly, the second harmonic of channel 3 mixes with channel 2 to produce a signal at 146.04 MHz, which is channel 4. Figure 4-13 shows that third-order mixing of channels 2 and 3 create signals on channels 1 and 4. If the receiver is tuned to channel 1 or 4 and strong channel 2 and 3 signals are present, then the nonlinearity of the RF amplifier or mixer will create internally generated interference on channel 1 or 4. This will limit the receiver's ability to detect a weak signal on channel 1 or 4.

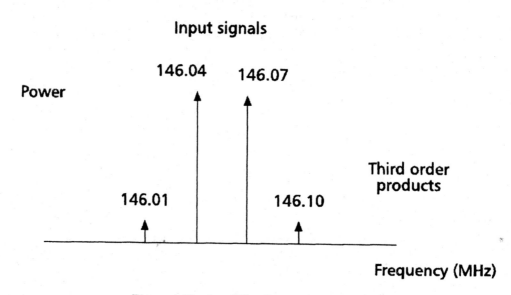

Figure 4-13: Amplifier Output in Figure 4-12

As the channel 2 and 3 signals grow stronger, more of the interfering channel 1 and 4 signals are generated. Figure 4-14 shows the relation between the power in the desired signal and the power in the third-order products. There is a linear relation between the output power and the input power of the desired signal until the device reaches saturation. At saturation the output power no longer increases for an increasing input. The apparent gain goes down. The maximum linear power output is defined as the output power where the gain is 1 dB less than for a small signal. This is referred to as the 1-dB compression point.

Third-order mixing products increase as the amplifier output increases. Due to the nature of the mixing, the power in the third-order products increases three times faster than the power of the desired signals. This can be seen in Figure. 4-14. Third-order products would eventually overwhelm the desired signal if the amplifier did not go into saturation. The point where the power in the third-order products is equal to the power of the desired signal is called the third-order intercept. The intercept point is a figure of merit because the amplifier will go into compression long before the intercept power is reached.

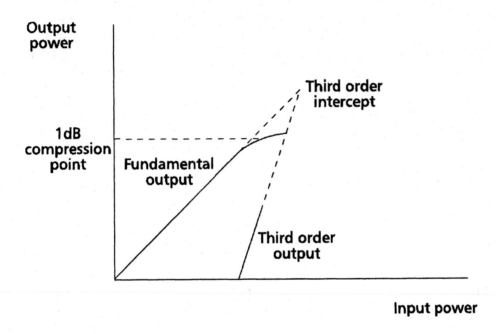

Figure 4-14: 1-dB Compression Point and Third-Order Intercept

The weakest signal the receiver can detect is limited by the internally generated noise. The equivalent power of this noise is the "noise floor." The strongest signals the receiver can tolerate are those that create a third-order mixing product equal to the noise floor. The ratio of the strongest signal to the noise floor is the spurious free dynamic range (SFDR).

$$SFDR = \frac{2}{3}(IIP3 - MDS)$$
(4.5.3)

where the noise floor is the minimum discernible signal, $MDS = kT_e B$ and $IIP3$ is the third-order input intercept point referred to the receiver input. High-gain low-noise RF amplifiers maximize sensitivity by reducing MDS. However, maximizing SFDR requires a careful trade-off between the gain, noise performance, and third-order intercept of all the stages prior to the narrowest filter in the system. Building a sensitive receiver is relatively easy; building a sensitive receiver that can handle strong signals can be an engineering challenge. Building a sensitive receiver that can handle strong signals with a minimum of battery power is more challenging still.

4.5.2.5 Selectivity

Filters play a key role in rejecting adjacent-channel signals, reducing the effect of the receiver's spurious responses such as the image, and reducing the effect of non-linear mixing products. The ideal filter would have a lossless passband centered on the desired frequency that is only wide enough to pass the desired modulation, and have zero response for all other frequencies.

Real filters have a passband with some minimal loss and then a transition band where the response falls to a specified level. This is the stopband, as shown in Figure 4-15. It is impossible to go from zero attenuation to some large attenuation with zero frequency offset. More complex filters, i.e., those having more poles, have more attenuation in the stopband and can transition from passband to stopband faster. However, this fast transition comes at the expense of waveform distortion in the time domain [Bli76]. This is of particular importance in systems employing digital modulation. Filter design and specification is both an art and a science [Zve67].

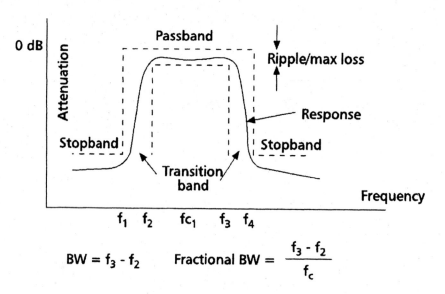

Figure 4-15: Typical Filter Response

An example of the bandwidth tradeoff can be seen in Figure 4-16. Using the fractional BW of Figure 4-15 the pre-selector filter is just over 1% wide. The pre-selector filter is chosen to do a credible job of rejecting spurious mixing products. A pre-selector filter that is much narrower will have too much loss. As bandwidth is reduced, filter loss tends to go up with an attendant increase in the NF. However, it is necessary to select only one channel of the example given in section 4.5.2.4. The IF filter must be approximately 30 kHz wide to pass the desired FM signal. However, the pre-selector filter cannot protect the RF amplifier and mixer from signals that may limit dynamic range.

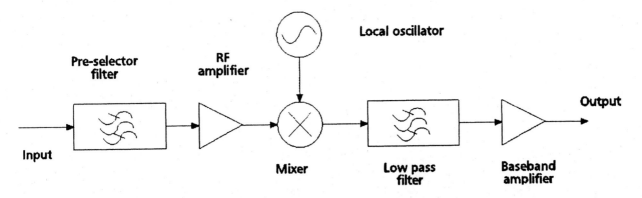

Figure 4-16: Relation Between the Receiver Pre-Selector Filter and IF Filter

4.5.2.6 Phase Noise

All superhet receivers have one or more internally generated signals or local oscillators. These oscillators are not noise free. The amplifier in any oscillator generates noise, which modulates the resulting oscillator output. Noise in oscillators takes two forms: amplitude and phase noise. The effect of amplitude noise can be neglected in well-designed oscillators [Roh88], but noise effectively phase-modulates the oscillator frequency. Oscillator phase-noise performance is measured in terms of the power spectral density of the resulting modulation sidebands. The spectral density can be predicted from the Q of the oscillator resonator and from the noise power generated by the active device in the oscillator [Rap02]. Figure 4-17 shows how the output spectrum of a typical oscillator is affected by phase noise.

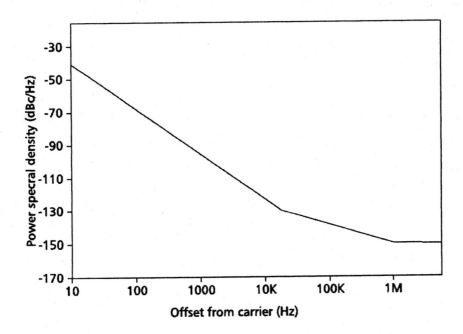

Figure 4-17: Oscillator Phase Noise

Phase noise degrades receiver dynamic range through what is called "reciprocal mixing." Figure 4-18 shows the effect of reciprocal mixing. The figure contains a noisy LO and a strong adjacent-channel signal outside the IF bandwidth. Without the LO noise, the IF filter would reject this signal.

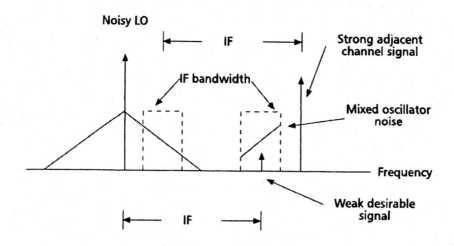

Figure 4-18: Reciprocal Mixing

The strong adjacent-channel signal acts as a local oscillator for the oscillator noise. This noise appears as a signal adjacent to the LO. Noise frequencies separated by the IF from the interfering signal will be mixed into the IF, as shown in Figure 4-18. The noise floor is increased by the oscillator phase noise, and this obscures the desired weak signal. In reality, all the signals in Figure 4-18 contain phase noise. Phase noise in the oscillator and in the strong adjacent channel will be mixed into the IF.

Phase noise produces additional receiver impairment in systems using phase-sensitive modulation, such as phase shift keying (PSK) or quadrature amplitude modulation (QAM). In digital PSK systems, it is necessary to regenerate the phase of the carrier signal as a demodulation reference. The phase noise of the receiver's LO is transferred to the incoming signal by mixing. This increases the phase uncertainty of the demodulation reference and leads to bit-decision errors. Phase noise has the effect of reducing the signal-to-noise ratio.

4.5.2.7 Direct-Conversion Receivers

The superhet receiver converts the incoming signal to an intermediate frequency. It is possible to frequency-convert the incoming signal directly to its original baseband. The LO is tuned to the frequency of the incoming signal. The mixer output is the sum and the difference of the signal and the LO frequencies. Because the signal and LO frequencies are equal, the difference is zero. The resulting output is the original information contained in modulation sidebands about the carrier frequency. this is called a direct-conversion (DC), homodyne, or zero-IF receiver. Figure 4-19 shows a block diagram of a direct-conversion receiver.

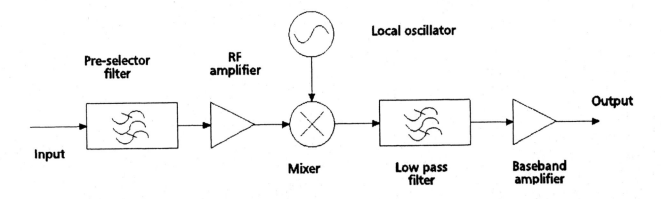

Figure 4-19: A Direct-Conversion, or Homodyne, Receiver

The DC receiver has the advantage of being simpler than the superhet. It is easier to build high-gain amplifiers at baseband than at the signal or IF frequency, so a DC receiver can be as sensitive as a superhet. This topology minimizes the number of components at RF frequencies In a simple DC receiver, the signal can be applied directly to the mixer.

Since all the signal processing occurs at baseband, it is possible to integrate the entire receiver including the mixer. This is attractive for small low-power receivers in devices such as pagers or in receivers employing digital signal processing (DSP).

A DC receiver is "tuned" to itself so LO leakage into the input is a problem. The oscillator must be carefully shielded to prevent this radiation. Oscillator phase noise increases close to the LO frequency, and this close-in oscillator phase noise is mixed directly into the baseband.

4.5.2.8 Other Receiver Topologies

Several other receiver topologies are of historical interest, although they find only occasional use. They are the tuned-radio-frequency (TRF), the regenerative, and the superregenerative receiver. The TRF topology consists of only the pre-selector filter, RF amplifier, and detector. There is no frequency translation. TRF receivers have been used in the LF and VLF range for standard time and navigation receivers. The received frequencies are generally fixed, and moderate-bandwidth filters can be made to match the modulation bandwidth. Because TRF receivers contain no mixing or frequency conversion, they can exactly reproduce the frequency or phase information in the original signal.

Regenerative/superregenerative receivers, which go back to the earliest days of radio, are capable of very high sensitivity with only a single active device. These receivers consist of an amplifier to which frequency-selective positive feedback is applied. External signals are applied to the amplifier and a signal close to the frequency of oscillation will be amplified repeatedly due to the positive feedback. The regenerative receiver can be thought of as a self-oscillating direct-conversion receiver. The single stage acts as a self-oscillating mixer.

Superregenerative receivers find application in short-range low-cost devices such as radio-controlled toys and garage-door openers. Their high sensitivity and extreme simplicity make them attractive for such applications, and they work well in the UHF range. However, they have poor selectivity, are easily overloaded, and they have the potential to create interference. Also, it is difficult to get stable, repeatable performance from receivers using regeneration.

4.5.2.9 Detection

Modulation is the process of attaching information to a carrier wave that is suitable for transmission through the channel, as shown in Figure 4-19. The modulator can vary the carrier amplitude, frequency, or phase. Equation 4.5.4 shows the quantities that can be varied in the process of modulation.

$$s(t) = A(t)\cos\{2\pi f_c(t)t + \phi(t)\} \tag{4.5.4}$$

Modulation can be analog or digital, and is widely covered in the literature [Cou93]. At the receiver, the process of demodulation, or detection, recovers the original information from the modulated carrier.

While demodulators vary, a commonly used demodulator is shown in Figure 4-20. This is a simple product detector that converts the signal down to baseband. Note that this is very similar to the DC receiver previously shown in Figure 4-19. The phase of the original signal must be recovered by additional circuitry. A more complex version of the detector is shown in Figure 4-21. An in-phase and quadrature baseband signal is generated and then digitized by the analog-to-digital converters (ADC). Digital signal processing is used to extract the modulating signal and the carrier phase. It is possible to reconstruct almost any signal with this demodulator. This type is widely used for complex modulations such as PSK, QAM, and orthogonal frequency division multiplexing (OFDM).

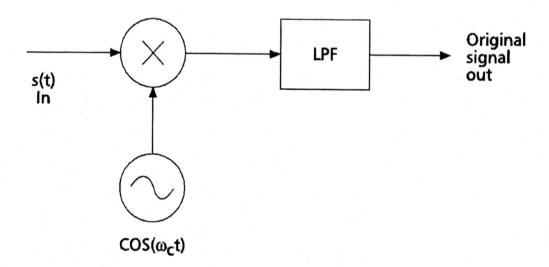

Figure 4-20: A Simple Product Detector

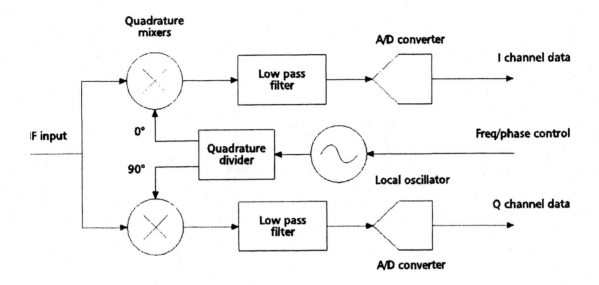

Figure 4-21: A Quadrature Detector

As noted in section 4.5.2.2, noise figure and effective temperature are used to measure receiver performance. While these measures are independent of modulation, they do not give direct insight into the noise performance of various modulation types.

For linear analog modulation such as large carrier AM (still widely used in broadcasting) and single sideband (SSB), the S/N_{in} or carrier-to-noise (C/N) ratio at the antenna is the same as the post detection or output S/N (S/N_{out}). Digital modulation performance is often specified in terms of E_b/N_0 (the energy per bit to noise density ratio) or E_s/N_0 (energy per symbol to noise density ratio). The noise density is the noise power in 1 Hz of equivalent bandwidth. For biphase shift keying (BPSK) where the carrier wave is multiplied by 1 for a one bit and -1 for a zero bit, the E_b/N_0 ratio is given by

$$\frac{E_b}{N_0} = \frac{CT_b}{kT_e} = \frac{C}{N} \tag{4.5.5}$$

where the bit period T_b is $T_b = 1/BW$. With E_b/N_0, it is possible to calculate the probability of a bit error or the bit-error rate (BER) for a particular S/N_{in} or (C/N) ratio. With higher order modulations where each phase or amplitude state represents more than one bit, E_b/N_0 becomes E_s/N_0, or the energy per symbol to noise density ratio.

Consider quadrature phase shift keying (QPSK) where there are four phase states and each phase state represents two bits. QPSK and BPSK have the same BER when E_b/N_0 is considered. However, QPSK has two bits per symbol so each bit receives half the power. QPSK requires 3 dB of additional carrier power to match the BER performance of BPSK. For BPSK, $E_b/N_o = C/N$, and for QPSK, $E_b/N_o = C/2N$.

QPSK is not a bad tradeoff because 3 dB in additional power results in 3 dB (X2) additional data that can be transmitted through a fixed bandwidth. Higher order modulations such as M-PSK, QAM, or OFDM do not provide this 1-for-1 increase in data throughput for a corresponding increase in power. They are not as power-efficient as QPSK. These higher order modulations are used where bandwidth is at a premium so the power disadvantage can be tolerated.

4.6 Key References

J.D. Kraus, *Antennas,* 3rd ed. ,McGraw-Hill, 2002.

T.S. Rappaport, *Wireless Communications,* 2nd ed. ; Prentice Hall, 2002

W. Hayward, *Introduction to Radio Frequency Design*, ARRL, 2004.

Chapter 5

Facilities Infrastructure

5.1   Introduction

This chapter outlines the information needed to specify, design, and implement wireless facilities and sites. The scope of this information is very broad, as can be seen from the list of topics below. Therefore, this chapter can only touch on these topics at a high level. At the end of the chapter there are references to books, standards documents, and web sites where detailed information can be found. Because wireless site design and construction, in particular, draw on many types of engineering backgrounds, information about possible sources of expertise in these areas is included throughout this chapter.

5.2   Contents

This chapter covers the following topics:

AC and DC Power Systems

Electrical Protection

Heating, Ventilation, and Air Conditioning

Equipment Racks, Rack Mounting Spaces, and Related Hardware

Waveguides and Transmission Lines

Tower Specifications and Standards

Distributed Antenna Systems and Base Station Hotels

Physical Security, Alarms, and Surveillance Systems

Industry Standards Bodies

5.3   AC and DC Power Systems

Communications equipment needs electrical power to operate. It is important to ensure that this power is properly provisioned. In order to determine what power will be needed to operate a site, it is necessary to understand how much power is required at the site and how long the site will need to operate should the normal energy supply be interrupted.

5.3.1   *Power System Specifics*

It is important to define the power/energy needs for the complete site. These include both the power consumption of the communications equipment and the power needed for support equipment supplying, for example, cooling or heating capabilities (see section 5.5). Both the maximum power consumption and the long term average power requirements must be calculated. The maximum power requirement must be known in order to size the site powering system, while the long-term average requirement must be known so that adequate long-term backup power or energy storage capacity is provided. Backup power needs to be evaluated and understood in terms how many hours of operation the back-up system must provide. Unique customer needs may require that back-up power be provided for an extended period of time.

Telecommunications systems are commonly powered from commercial utility AC power that has been converted to −48VDC or ±24VDC. In addition to the energy conversion equipment, energy storage must be provided on-site to maintain the operation of the equipment in the event of a loss of utility power. The most common energy storage system is a bank of batteries; however, flywheels and other new energy storage technologies have been proposed for telecommunications applications. A second level of backup power may also be required and should be considered. For large installations, a diesel generator, turbine alternator, or other backup power source may be required. Any such source must include an automatic start capability in the event of a loss of commercial power, and, of course, a source or supply of fuel (gasoline, natural gas, etc.) must be provided at the site.

There are other options but the most common power plant for wireless base stations consists of centralized AC-to-DC rectifiers and battery storage, with DC distribution conductors carrying the power at the appropriate voltages to the communications equipment. Figure 5-1 shows a block diagram of a typical power system. Since most solid-state devices and circuits operate at voltages other than −48 VDC and ±24 VDC, internal power supplies within the communications equipment will typically include DC-to-DC converters which step these voltages down to the levels used by the electronic circuits. In some cases, such converters may need to be provided as separate items of equipment to handle the cumulative low-voltage power needs of multiple pieces of communications equipment.

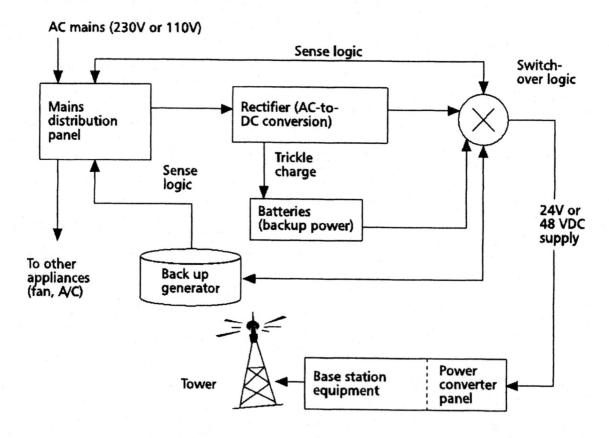

Figure 5-1: Typical Wireless Base Station Power System Layout

In addition to the DC power for the communications equipment, AC power must be distributed throughout the facility or site to power air conditioners, fans (although typically not the cooling fans that are incorporated into individual pieces of equipment), electrical heaters, and lighting, as well as to provide accessible outlets for powering test sets or other maintenance and repair tools. Whereas the DC voltages are not hazardous (but the current levels can be!), AC voltages can be hazardous. Therefore, industry standards, practices, and local codes all must be adhered to in the design and construction of these AC distribution networks. Proper shielding and grounding are also necessary to avoid introducing AC "hum" into any communications path.

In remote locations, alternative energy sources are being utilized more and more often. These include solar (photovoltaic) panels, wind turbines, fuel cells, and other new technologies. Because solar and wind energy are by their nature intermittent, the use of these energy sources requires a large battery bank or other energy storage facility on site to maintain the site operation in the event of long periods of cloudy weather or the absence of wind. Fuel cells, while not intermittent, require a supply of fuel, which must be replenished as it is consumed during the cell's operation. Whatever alternative energy source is used, the system must be sized, designed, and constructed to deliver both DC and AC power to meet the needs of both the communications equipment and it's support equipment.

5.3.2 Additional Resources

A great deal of detail is involved in the design and implementation of AC and DC power systems. Because these systems usually connect to a commercial power company and involve potentially dangerous voltages and currents, they must be designed with care and they must conform to local codes. Because of these issues of safety and conformance, it is good practice to involve specialists with expertise in power system design and in locating the appropriate codes and standards that the power system must meet. Section 5.12 identifies some of the resources available, and suggests ways to locate such relevant expertise.

5.4   Electrical Protection

Engineered protection schemes are required at communications facilities to mitigate the effects of transient voltages. These transients are the result of induced and conducted lightning currents, as well as power circuit switching operations. The latter include switching of power factor correction capacitors on the medium-voltage AC distribution system, along with any other make or break operation in a power system not occurring at a zero current point, such as motor contactors, fuse or circuit-breaker tripping, etc. Whenever an energized circuit is opened, the collapse of the magnetic field self-induces a transient voltage surge into the conductor.

5.4.1 Lightning Strikes as Sources of Transients

The largest protection threat to a communications facility is a direct lightning strike to the radio tower. During a tower strike, all metallic entrances to the facility (AC, Telco, water/gas lines, etc.) are used as lightning-current exit paths. This is due to the high ground potential rise (GPR), also known as earth potential rise (EPR), of the facility ground reference with respect to remote earth.

Figure 5-2 illustrates typical site GPR/EPR during a lightning strike. It assumes a nominal 20-kA peak lightning current and a 5-ohm ground resistance. The figure of 20 kA is a mean value for lightning currents; approximately 50% are lower and 50% are higher. Lightning originates from both a high source voltage and source impedance. The facility ground resistance is negligible compared to the source impedance and has little effect on the delivered lightning current. Because of this, lightning is considered a constant current source. As shown, a nominal 20-kA strike through a nominal 5-ohm ground results in the facility ground plane rising 100 kV above remote earth.

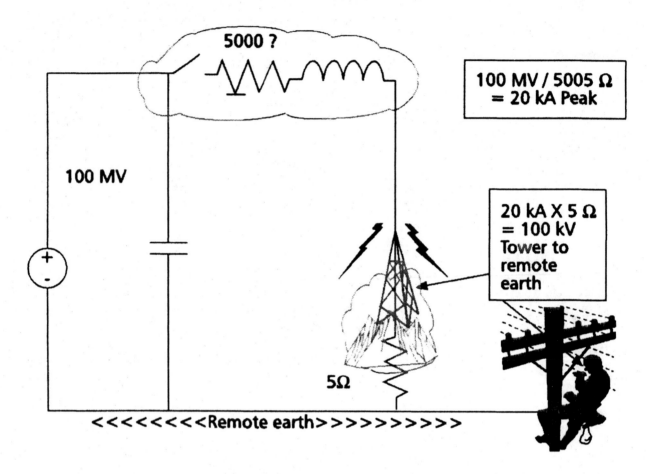

5000 ?

100 MV / 5005 Ω = 20 kA Peak

100 MV

20 kA X 5 Ω = 100 kV Tower to remote earth

5Ω

<<<<<<<<Remote earth>>>>>>>>>>

Figure 5-2: Ground/Earth Potential Rise Due to a 20-kA Lightning Strike

All metallic objects, including security fences and gates, should be solidly bonded to the facility ground reference. As a general rule, any two metallic objects within a person's arm-span should be bonded to the common facility ground reference for safety.

5.4.2 Surge Protective Devices

Non-bonded metallic conductors, such as AC power and telecommunications cables, should be indirectly bonded to the facility ground through surge protective devices (SPDs). Since these conductors are referenced to remote earth, the high GPR/EPR will force their use as exit paths. The use of SPDs will prevent external flashover and will hold the incoming conductors close enough in potential to the facility ground reference to prevent equipment damage. AC SPDs (also called Transient Voltage Surge Suppressors [TVSSs]) are labeled with a maximum available fault current rating and what, if any, upstream over-current protection device is required. Care must be taken to ensure that the anticipated fault current at the point of SPD installation does not exceed the labeled rating on the SPD, and that any required upstream over-current protection is present. All SPDs used on communications facilities should be listed by a nationally recognized testing laboratory (NRTL) and used in accordance with the electrical code requirements. In the U.S., the National Electric Code (NFPA-70, also known as NEC) article 285 covers SPDs for AC power circuits.

European Council Directives (EEC) mandate that products offered for sale or placed in service in Europe must comply with the Member State's requirements as prescribed in all relevant CE marking directives. Related to specific topics, CE marking directives require EU Member States to appoint Notified (Type A Inspection Body) and Competent Bodies to evaluate and approve products against the requirements identified in each directive. Such bodies must be accredited against the essential requirements of EN 45004 and EN 45011, and the European Commission notified of their appointment. There must be demonstrable independence between a Notified Body and any activity associated with the design, manufacture, supply, installation, purchase, ownership, use, or maintenance of the items inspected. Essentially the Low Voltage Directive applies to consumer and capital electrical equipment (mains energized electrical appliances) designed for use within the voltage ranges from 50 VAC to 1,000 VAC and from 75 VDC to 1500 VDC. However, equipment subject to RTTE directive 1999/5/EC must meet the requirements of the Low Voltage Directive with no lower voltage limit. For example, in the UK, CE marking certification is performed by such private laboratories as Mira Ltd.

Elsewhere, country-specific codes and requirements apply. For example, in Japan the applicable requirements are contained in the Electrical Appliance and Material Safety Law (DENAN) (formerly the Electrical Appliance and Material Control Law (DENTORI). In Malaysia, the national testing laboratory is SIRIM, and its electrical testing typically follows standards created and issued by the International Electrotechnical Commission (IEC).

All well-designed and constructed brands of AC SPDs utilize the same basic components and have approximately the same let-through voltage. The primary difference in AC SPD performance is not brand, but installation. AC SPDs are normally connected in parallel with the power circuit. The parallel or shunt lead length used to connect the AC SPD can significantly degrade performance. This is due to the inductive voltage drop on the leads at lightning frequencies. For example, a typical SPD for 120-VAC service has a let-through voltage of approximately 500 V on a standard 10-kA 8/20-µs surge. Closely coupled connection leads, on the same surge, add an additional 600 V/m (200 V/foot). Widely separated leads have an even higher voltage drop. For acceptable installed performance, lead lengths should be kept as short and straight as practical, and as a rule of thumb should not exceed 1m (three feet). Outside North America, where 230 VAC is the common mains voltage, it is essential to follow practices based on applicable local standards. Several standards and work practices for safety in electrical testing are published by British standards bodies for work with low-voltage electrical systems of up to 1,000 VAC; similar publications are available in other countries and other regions and should be consulted when planning the AC protection system for any site.

SPDs for communications circuits are addressed in a variety of documents. In the U.S., the applicable requirements are included in NEC 800c; in France, AFNOR French Standard UTE C15-433 covers such SPDs. There are multiple categories and listing requirements for communications SPDs. Different systems, installation topologies, and wiring methods have their own unique characteristics and require different protector performance to mitigate risks and operate without introducing hazards. Choosing the wrong protector reduces (or eliminates) the effectiveness of the protection scheme and increases the risk of catastrophic protector failure, electric shock, fire, and injury.

For circuits directly connected to outside plant (OSP) conductors, the protector must protect against both lighting and accidental contact of the OSP conductors with joint-use AC distribution conductors (power-cross). In the U.S., the NEC requires use of a listed primary telecommunications protector on exposed OSP circuits. These devices are NRTL-listed in the U.S. to Underwriters Laboratories UL497. Outside North America, where 230 VAC is the normal mains power, local jurisdictional requirements and codes should be followed to ensure proper protection in the event of a power-cross. The primary protector is normally supplied by the telecommunications service provider. Questions regarding the provided communications circuit classification should be addressed to the electrical protection engineer employed by this provider.

When a protector is directly connected to the metallic conductors between the listed primary protector and the first piece of electronics equipment, it is classified by the NEC as a secondary protector. Although the U.S. NEC does not require use of secondary protection, it does require that if used, a secondary protector must be listed for the purpose. These devices are NRTL-listed in the U.S. to Underwriters Laboratories UL497A, which requires in-line fusing to prevent wiring fires during power-cross events. This fusing limits the surge capability of the secondary protector to less than that required for primary protection. Secondary protectors should only be used downstream of a listed primary protector. In other countries, local jurisdictional codes apply; see for example the British Standard "Customer-Owned Outside Plant (CO-OSP) Design Manual, Third Edition".

Communications circuits that are not exposed to OSP surges or power-cross, are classified as isolated loops. Isolated loops include intra-site alarm and communications circuits, as well as many OSP-derived circuits. Often, the telecommunications service provider uses active listed electronics to convert from OSP circuits (such as DSL) to internal circuits (such as T1/E1 or Ethernet). In this case, the listed conversion electronics isolate the site's internal circuits from OSP surges and power-cross, and the internal circuits are classified as isolated loops.

Isolated loop protectors are NRTL listed in the U.S. to UL497B. Listing of an isolated loop protector does not require survival under the large GPR/EPR surges experienced during lightning strikes to the facility. Because of this, it is important to evaluate the surge capability of all isolated loop protectors used on communications facilities. Protectors at communications facilities rated for a multiple-pulse maximum-surge-current capability of 10 kA, 8/20 µs have proven robust enough to provide reliable service.

At the time of this writing, UL is in the process of developing additional UL497 family specifications for isolated loop protectors subject to GPR surges, and for RF antenna surge protectors. Persons responsible for wireless site design and construction should remain cognizant of such developments in whatever local jurisdictional standards apply to the site being planned.

5.4.3 Surge Threat Evaluation

A tool commonly used in the U.S. to make lightning-risk site-selection decisions, or to determine if a given problem was lightning related, is the data from the U.S. National Lightning Detection Network (NLDN). The NLDN collects data on over 20 million lightning strikes to the continental U.S. each year, including time, longitude, latitude, and peak current. Data sets for a particular location and time are available online for a nominal charge at http://thunderstorm.vaisala.com/.

5.4.4 Additional Resources

Another protection resource available in the U.S. is the Protection Engineers Group (PEG). PEG is an ATIS-sponsored organization of protection engineers specializing in the electrical protection of communications facilities. For twenty years, PEG has held an annual conference where noted industry experts discuss communications facility protection issues. At the time of this writing, PEG is working to create CDs for purchase on topics from past conference presentations, as well as future Webinars with communications facility protection topics covered by noted industry experts. Information about PEG can be found at http://www.atis.org/peg/.

5.5 Heating, Ventilation, and Air Conditioning

Communications equipment typically requires a controlled environment to operate reliably. For some equipment the temperature—and possibly the humidity, airflow, or air quality—must be tightly controlled. For other equipment, the environmental conditions can be allowed to vary over a wider range. It is important to understand the applicable requirements and to ensure that appropriate solutions are provided.

5.5.1 Heating, Ventilation, and Air Conditioning (HVAC) Requirements

Because communications equipment consumes power and dissipates heat, most such equipment requires some degree of environmental cooling to maintain its internal temperature within a range where it will operate reliably. In extreme environmental conditions, heating may be required to similarly maintain equipment at minimum operating temperatures. Batteries are commonly used as a backup power source and are typically quite sensitive to temperature, both as regards their energy storage capacity and their service life. Adequate backup power, including alternate power sources, may be required to maintain the controlled site environment if the potential site temperature can reach hot or cold extremes.

Storage batteries also require ventilation. Hydrogen and oxygen gases are generated as a function of overcharging. The amounts of both gases are directly proportional to the current that flows through the battery cells and the number of cells in series. There have been numerous events in which hydrogen gas explosions have damaged or even destroyed a communications facility. In most of the IEEE standards for stationary batteries, the maximum allowable gas concentration is 2% in order to provide a margin of safety against the lower flammability limit of hydrogen, which occurs at a concentration of 4% or above.

5.5.2 Additional Resources

When buildings are involved, it is often advisable to work with a mechanical engineer experienced in building systems to calculate the ventilation system requirements. An engineer with such experience can also provide information on the types and capacities of any cooling or heating equipment, based on the energy consumption and dissipation of the communications equipment that is to be located within the building.

5.6   Equipment Racks, Rack Mounting Spaces, and Related Hardware

Standards have been established to mount various types of equipment in common frameworks. Typically, these were originally developed many years ago by the telephone and broadcast industries and have subsequently migrated into all areas of electronic and industrial systems. A basic understanding of these standards, and where to find them, is essential to ensure that the site can accommodate the planned equipment, and that the equipment's needs for power, ventilation, technician access, etc., will be met. Complete site equipment requirements must be determined to ensure that there are no spacing issues. Both horizontal and vertical space requirements must be clearly defined at the outset. A helpful way of checking for spacing issues is to generate a complete set of blueprints for the proposed equipment layout.

5.6.1 Mounting Standards

The primary standard in the United States is included in ANSI/EIA/TIA 310 and applies to equipment that mounts on a 48.3-cm (19″)-wide rack. All equipment that mounts in such racks has a front panel 48.3 cm (19″) wide and is sized in increments of 4.4 cm (1-3/4 in.) high.

In telephone central offices and similar facilities, there is a comparable standard in which the front panels are 58.4 cm (23″) wide and sized in increments of 5.1 cm (2 in.) tall. This system also allows the equipment height to be sized in factional units, i.e., in whole- and half-inch increments. Details on this standard rack format can be found in the catalogs of most providers of equipment racks for the telecommunications industry.

Both of these standards also include information on such things as the types of mounting hardware—for example, the diameter and thread pitch of the mounting screws for the equipment. There are also details about the installation of the racks themselves—for example, the size and spacing of the bolts by which the racks are anchored to the floor. Additional information applicable to a specific installation may include provisions for overhead bracing and for troughs or other guideways for containing and routing interconnect wiring and cable. In areas subject to significant earthquake risks, additional bracing, suspension from overhead beams, or other special design features may be required.

Because operating communications equipment generates heat and requires a flow of cooling air to maintain reliable operation, individual pieces of equipment may require space for heat baffles. Angled heat deflectors are often inserted between pieces of equipment. These typically direct the heat flow from the lower piece of equipment out toward the back of the rack, while simultaneously providing space for cooler air to flow in from the front of the rack and up into and through the higher piece of equipment. The space occupied by such heat deflectors must be planned for when calculating the amount of equipment to be installed in a rack.

While individual racks can be mounted abutting side by side in rows, adequate space must be allowed between rows. In the front of equipment racks, such space is obviously required to allow technicians to access the equipment for test, maintenance, repair, and replacement purposes. However, adequate space must also be allowed behind racks, since this is where most connections are made—to the power distribution bus, to the waveguide or cabling leading to the antennas, and to the cables or other communications connections that interconnect pieces of equipment, and that connect the site to the ground-based telecommunications network.

5.6.2 Additional Resources

Unlike power distribution or electrical protection, there are few specialists in rack design and mounting. It is therefore important to understand the general nature of the equipment and its physical requirements for access and heat dissipation. Often there are also codes and regulations that define the access space that must be provided for personnel. All these factors must be considered when designing a facility, and it is advisable to review the equipment manufacturers' literature in some detail to understand how best to mount it and provide the necessary access.

5.7 Waveguides and Transmission Lines

Waveguides and coaxial transmission lines are the most common devices for connecting transmitters and receivers to antennas. In this section we address some of these devices' physical characteristics that need to be understood in planning the layout of a wireless facility.

5.7.1 Physical Characteristics of Waveguides and Coaxial Cables

Waveguide today is mostly elliptical and is handled much as a cable. The minimum bend radius is a very critical factor and is relatively large with respect to waveguide dimensions. Those dimensions and specific handling recommendations are available from the waveguide manufacturer. Strict adherence to these requirements is critical to the waveguide meeting its performance characteristics.

Coaxial cables are similar but not as critical as waveguide with regard to careful handling. However, strict adherence to the manufacturer's requirements is critical to meeting the specified performance characteristics. In general, coaxial cables have more transmission loss compared to a waveguide. As a rule of thumb, the more flexible the cable, the greater the potential loss.

Cables and waveguides must be supported at least every 1m (3 ft) on the horizontal and vertical when installed. Terminations may require short twistable-flexible waveguide sections at one or both ends to accommodate sharp angle bends that may be necessary to get to equipment or antennas. Large-diameter coaxial cables chosen to reduce transmission losses may require short sections of small-diameter highly flexible cables at the ends for the same reason.

Special provisions must be made when a waveguide or coaxial cable is to enter a building or equipment housing from outdoors. Such entry may require a bulkhead fitting or other feed-through device. The waveguides and coaxial cables must generally be bonded to ground prior to entering a building or equipment housing (see section 5.4). Special fittings are generally available from the manufacturer for this task.

The maintenance of waveguides and coaxial cables is also critical to their ongoing performance as a link in a communications system. Dry air is often pumped into a waveguide to keep moisture away so as to reduce losses and resist deterioration of the waveguide's electrical performance. Pressurized waveguide needs to be checked periodically to determine if the system is holding pressure; once a year is the minimum suggested frequency. Overall system testing and inspection should also be done on an annual basis to identify and, if possible, prevent deterioration of the system.

5.7.2 Additional Resources

The most useful reference documents on this subject are usually the manufacturers' literature. They generally provide comprehensive guides to handling, terminating, mounting, and grounding the waveguide or coax.

5.8   Tower Specifications and Standards

Wireless communications towers are usually designed or selected by a professional structural engineer. However, the design depends on a complete understanding of the electrical and physical characteristics needed for the wireless system to perform properly. Tower specifications are also required to guide the fabrication and installation of any tower that will support communications antennas.

5.8.1 Tower Requirements

Towers at wireless base stations not only support the antennas; they must support them in the proper positions and orientations to achieve the desired coverage, and they must do so under wide-ranging environmental conditions. Wind, rain, and snow and ice loads apply mechanical stresses to the tower. Dust, lightning, and salt fog (near the ocean) can erode or damage the tower, and the resulting corrosion can degrade its strength. Uneven heating due to solar radiation can cause the tower to flex, altering the antenna orientation and adding stresses to the joints in the tower structure. Towers must be robust enough to carry all imposed loads (antennas, equipment, and environmental) without significant distortion, and must do so for many years, or even decades, of service.

The tower design also must accommodate potential future loads. The full capacity tower loading must be analyzed to ensure that the tower is not overloaded beyond the physical weight it can withstand. This should include all future antennas that may be added, along with their associated cabling or waveguides and the potential ice, wind, or other loads that these antennas and cables/waveguides may impose. It is very common to reuse a good site, but often times the tower may already be overloaded. A new tower may need to replace the old one or to be erected nearby to supplement the older one. It is therefore good practice—and often cost-effective in the long run—to consider future needs and over-engineer the tower for today.

Tower designs are specified based upon the type of tower structure required for the specific project, for example, monopole, three-legged, or four-legged lattice structures. Before a tower can be designed, the total antenna load needs to be determined. The design of the structure depends upon the loading of both the antennas and transmission lines to be placed on the tower, as well as the wind design load. In addition, towers in areas where ice and snow occur need to specify a number designating the ice loading that needs to be considered.

Proper tower design and installation also require a detailed geological study to determine the type of foundation necessary to support the structure. If the natural soil at the site is inadequate to support the specified fully loaded tower, materials may need to be brought to the site to strengthen the soil to the required standards; alternatively, pilings or other supports may be required to distribute the tower load and transfer it to bedrock or other sufficiently strong underlayment. In addition, specific local requirements may apply, and these need to be determined, along with local codes and planning constraints.

5.8.2 *Additional Resources*

ANSI/TIA 222-G, "Structural Standards for Steel Antenna Towers and Antenna Supporting Structures," is the document used in North America to specify towers. It is incorporated in most building codes by reference in the Structural Design section. This standard is continually being reviewed by experts in the design and erection of towers. The American Society of Civil Engineers (ASCE) is another good source for standards related to building and structure load requirements.

Because of the critical role that the tower plays in the performance of a wireless site, it is advisable to seek out the expertise of a structural engineer who will understand the behavior of a tower under the applied loads, as well as the applicability of various codes and standards in the selection of the tower design. Often the local power utility or similar institution may be able to provide such services.

5.9   Distributed Antenna Systems and Base Station Hotels

Due to the dense population in urban areas, it is often necessary for carriers and service providers to use multiple towers throughout a city. However, locating such towers in heavily built-up areas can be very difficult. And even with multiple towers, those inside buildings often do not get strong, reliable signals. Therefore, an alternative approach to the traditional tower/antenna infrastructure has been developed in recent years in which many urban structures are served by a Distributed Antenna System (DAS) that is housed in a Base Station Hotel (BSH).

5.9.1 *In-Building Facilities Infrastructure*

The BSH is an indoor area, typically in the basement or near the roof of a building where there is enough space for all the equipment at the site. The BSH serves as an alternative means to the conventional tower/hut infrastructure for providing indoor coverage. A centralized BSH uses DASs installed inside nearby buildings for indoor distribution of the RF signals. Figure 5-3 shows the basic elements of a BSH/DAS system. Typically, a BSH consists of base station equipment from different service providers connected through a multi-channel amplifier. All incoming signals are combined onto a single RF carrier that is usually converted into an optical signal carried over a single-mode fiber. The entire system consisting of multi-channel amplifiers and the RF-to-optical converter is known as the Master node.

Fibers from the Master node are then distributed to different buildings in the area. Each building has receiving nodes known as the slave nodes that convert the signals back from optical to RF. The RF signal is then radiated within the building using multiple wall and/or ceiling mounted antennas that serve different rooms and areas. Typically, each master node can support 24 to 32 slave nodes. The DAS system is quite common in many cities today and serves a variety of urban buildings ranging from airports and subway systems to office buildings and underground passageways that connect large malls. Unlike tower-based systems, DAS systems do not radiate signals over an open geographic area. Instead, they are designed to provide only local coverage within office buildings or restricted spaces where signals from towers clearly cannot reach or provide adequate signal strength. In terms of layout, the BSH is not very different from the base station layout described in Figure 5-1. The major difference is that the antenna tower does not exist, and, therefore, antenna diversity schemes cannot be used. Also, the output from the base station equipment is not carried as RF but is converted to optical for easier transport and reduced transmission loss over significant distances in urban areas. Service providers typically connect their base stations located in a BSH to their network using traditional backhaul methods such as T1/E1.

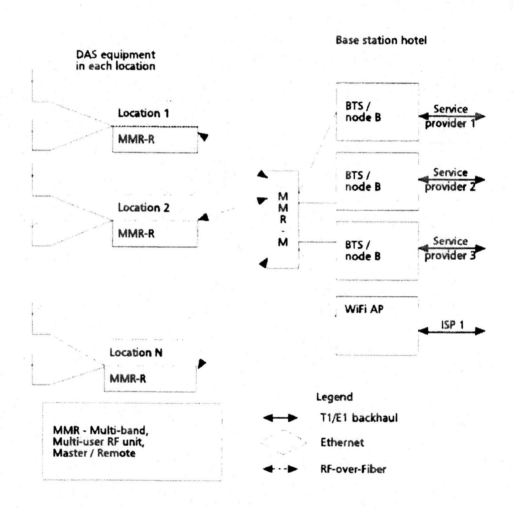

Figure 5-3: Example of a Distributed Antenna System and Base Station Hotel

5.9.2 Additional Resources

Since both the BSH and the DASs are located within buildings, the most useful resource is often the appropriate section of the local building code. Also, because it is often easier to obtain space in a new building, or one undergoing extensive renovation, local architects and contractors can be helpful sources of information. This is particularly true when reliable wireless access is being promoted as a feature of the new or renovated building as a way to attract tenants.

5.10 Physical Security, Alarm and Surveillance Systems

Telecommunications facilities are often critical infrastructure and as such are potential targets for a variety of threats, natural or man-made. The degree of protection needed is a function of the criticality of the facility in question.

5.10.1 Importance of the Facility

A network plan generally will indicate the importance of a facility. A cell site that is one of many in an overlapping network is unlikely to be as critical as a single remote communications vault on a mountain top that covers 500 square miles and may include public safety radio facilities. For sites that will carry high-priority traffic (e.g., security or public safety communications), complete site redundancy should be considered. Power backup and even equipment backup or redundancy will be useless if the entire physical site is lost or destroyed, whether due to a natural (e.g., a severe storm) or man-made (e.g., arson or other attack) catastrophe.

5.10.2 Team Effort

The security of a facility is normally a team effort in most organizations. The organization generally has a security group whose responsibility is the overall security of all of the organization's assets. The systems implemented to protect the communications facilities should interoperate with the general security plan of the organization.

5.10.3 Physical Security Considerations

Many issues must be considered when locating a facility. Flood, earthquake, fire hazard, and potential vandalism are just some factors that affect physical security. Remote sites in forests or wilderness areas should be protected from potential damage by local wildlife. In today's world, the possibility of terrorist attacks should be considered when locating and constructing critical communications facilities.

5.10.4 Alarm Systems

Facilities should be provided with a remotely monitored alarm and surveillance system. Such a system should remotely report the status of all equipment and infrastructure at the facility. This should include unexpected entry to the facility, as well as temperature, humidity, and fire alarms. When sufficient bandwidth is available, video surveillance may be included.

5.10.5 Electronic Access System

Where electronic access systems are required, it is necessary to provide the bandwidth needed to implement site access authorization. The associated equipment should be powered via a battery backup system. This includes the actual door access devices as well as the access electronics. A secure backup plan should be provided to allow access in the event of the failure of the electronic access system.

5.10.6 Additional Resources

Site security and alarm systems are often overlooked in the process of planning a facility yet are important issues that must be addressed. A helpful resource may be the site specifications for other facilities maintained by the communications company, especially those in similar locations. Contact with the company's security organization is also recommended. Also of help is the literature from the manufacturers of various types of surveillance and monitoring systems whose capabilities are most applicable to the site.

5.11   National and International Standards and Specifications

Industry standard specifications and national and international standards make the job of site design and construction a bit easier. Many of the best practices make systems more reliable by drawing on the experience of others. In many cases the application of proven good practice helps speed approvals by local jurisdictions and may avoid unnecessary government regulation.

5.11.1 Industry Standards Bodies

ATIS – The Alliance of Telecommunications Industry Solutions (ATIS) is a US-based organization that is committed to rapidly developing and promoting technical and operations standards for the communications and related information technologies industry worldwide using a pragmatic, flexible, and open approach.

CDG – The CDMA Development Group is a non-profit body that developed and has continued to evolve CDMA technology since 1993. All major manufacturers of both handsets and infrastructure equipment, as well as wireless service providers, are members and work to foster development of CDMA technology . Working closely with 3GPP2, CDG drives the evolution of CDMA standards worldwide.

ETSI – The European Telecommunications Standards Institute (ETSI) is an independent, not-for-profit standardization organization of telecommunications organizations in Europe, with worldwide projection.

GSMA – The GSM Association is a nonprofit body that focuses on TDMA-based technology. GSM has been a widely used worldwide standard for 20 years; GSM standards have a significant focus on the user interface in the handset as well on the interfaces between elements of infrastructure equipment. Very detailed standards (running into 12 volumes) were evolved that resulted in a true plug-and-play approach for equipment from different vendors manufactured to GSM, GPRS, and EDGE standards. GSMA works closely with 3GPP to drive the evolution of W-CDMA standards worldwide.
. GSMA works closely with 3GPP to drive the evolution of W-CDMA standards worldwide.

IEC – The International Electrotechnical Commission (IEC) is a not-for-profit, non-governmental international standards development organization that prepares and publishes international standards for all electrical, electronic, and related technologies.

Telcordia – In the United States, Telcordia publishes Generic Requirements documents that are used by many service providers as the basis for the procurement of reliable and standardized equipment. Consequently, many manufacturers also test their products to these requirements, which often function as de facto performance standards. Among Telcordia's documents, the New Equipment and Building Standards (NEBS) have wide applicability to many of the topics related to the communications network infrastructure, including wireless site installations. NEBS addresses the physical aspects of the equipment installation, including test specifications for shock, vibration, temperature, etc.

3GPP – The European-based 3rd Generation Partnership Project is an international body that publishes wireless standards for W-CDMA as the successor to GSM. All wireless equipment worldwide (including handsets) are bound by this standard. 3GPP is exclusive to the wireless industry and addresses all aspects of wireless infrastructure such as base stations, Mobile Switching Centers, base station controllers, and handsets. The North American counterpart known as 3GPP2 addresses wireless standards based on the CDMA 2000 standard. 3GPP2 is similar in structure to 3GPP but is limited to the North American version of the CDMA standard. All major manufacturers of both handsets and infrastructure equipment, as well as wireless service providers, are members of these standards bodies. It is important to note that the wireless equipment located in any base station area is governed by one of these two major international standards bodies.

5.12   References

The design of a wireless site requires the specification of numerous site components in addition to the communications equipment. These include the power system (primary and backup), equipment required to provide heating, cooling, and ventilation for the site, the tower, and in some cases, even the building or hut in which the communications equipment will be housed. The specification of such a broad range of components involves knowledge of many disciplines, applicable standards, and local codes. For this reason, it is advisable to seek expert assistance from others with experience in these areas.

5.12.1 Locating Professional Assistance

Many of the professionals who can assist in the design of the site infrastructure (power, HVAC, on-site fabrication, etc.) are connected to the building construction industry. Architects and engineers often are already known by the organizations involved. They are generally familiar with specialists in areas associated with communications facilities.

Local or national engineering associations are also good sources of contacts. Websites can help locate the local chapter or section of a professional association. The practical experience of workers in the applicable construction trades (for example, electricians) can be helpful, and they often may be familiar with the reputation and past performance of local architects and design engineers. Involving local experts can also be helpful in understanding any unique aspects of local codes, regulations, site inspections, and the permitting and approval process.

Chapter 6

Agreements, Standards, Policies, and Regulations

6.1 Introduction

Agreements, Standards, Policies and Regulations (ASPR) are the necessary mechanisms of the wireless industry used to anticipate, improve and control the behaviors of entities that design, develop, implement, operate and evolve wireless communications networks [ATIS07]. ASPR, sometimes abbreviated as "Policy", is one of the eight fundamental ingredients that make up the wireless communications infrastructure [Rau06, CQR07, NRIC05]. Similar to the critical role of hardware in providing the physical medium for electronic activity, and the critical role of software in providing the logic, automation, and control that shape communications features and services, ASPR plays a critical role in coordinating a wireless network's many diverse elements and stakeholders [NRSC02, NRIC03, NSTAC06]. Without this coordination, equipment from different suppliers would fail to interoperate; interfaces between network operators would be incompatible; the inconsistency of subscribers' experiences would be intolerable; and the result would be far less useful networks.

ASPR also help ensure that wireless networks do not disturb their environment or other interests. In addition, they provide a framework where fairness can be managed among competitors. However, the primary reason for ASPR is to promote the reliability and quality of the wireless user's experience and the interoperability of networks. As will be seen, most implementations of ASPR do not require government involvement. Today's commercial legal frameworks are sufficient to support most of the coordination needed.

6.2 Contents

This chapter addresses the following topics:

Agreements

Standards

Policies

Regulations

6.3   Agreements

6.3.1 Background

Agreements are mutually accepted terms that define the expectations among two or more parties. In the wireless communications industry, agreements address issues such as a network operator setting expectations for product quality with its equipment suppliers; legal arrangements between competitors to share network resources during a time of crisis; and service agreements with consumers regarding such things as network access coverage, pricing, and the types of services provided.

Agreements do not require government involvement, but rely on commercial legal frameworks.

Because the agreements are within the control of the parties involved, they are a relatively fast means of controlling the parties' behavior.

6.3.2 Importance of Agreements

Agreements are needed because the stakeholders involved in wireless networks have different interests. The agreements provide for tradeoffs so that each party's interests can be protected in an acceptable way.

Consumers are the most numerous stakeholders in the industry, their number being approximately a billion globally. Their interests vary around the common points of price, features and quality. Service agreements are usually implemented in service contracts, which typically have options.

Service providers are the stakeholders closest to the consumer in the supply line. Some service providers do not own network equipment, relying instead on network operators through legal agreements to sell services that are run on networks operated by the latter. Other service providers are also network operators, which is the next role further back in the supply line. Network operators design, build, operate, and maintain networks. Typically, they have agreements to provide roaming services to the other company's customers.

An industry role that parallels the network operator is the facilities operator. Facilities operators include those that manage cell sites and other "telecom hotels." These are co-location facilities, where multiple network operators share a common physical site for reasons that include reduced cost, local restrictions, or the need to interconnect.

As mentioned above, network operators also have agreements with their equipment suppliers. Agreements can cover the delivery schedule of new features and technical capabilities, the quality and reliability of products; the behavior of individuals who will be on-site during equipment installation or upgrade; and penalties for failure in any of these areas.

6.3.3 Examples of Agreements

The following examples provide insight into the nature of agreements.

Example 1. Service provider-consumer agreements

One of the most common agreements is between a consumer and a wireless service provider. A simplified generic example of such an agreement is shown in Figure 6-1, which shows a trio of service options corresponding to different prices.

	Option A	Option B	Option C
Coverage	Hometown region	National	International
Services	Voice	Voice, text	Voice, text, video
Minutes per month	100	1000	Unlimited
Price	25	100	250

Figure 6-1: Examples of Options in a Consumer Service Agreement

Example 2. Mutual aid agreements

Competitors can reach an agreement well in advance of a crisis to offer equipment or other resources when human life is at stake. A wireless network operator can greatly increase its network resilience by setting up a legal framework in which an entity that may be a competitor on any other day would, during a crisis, make available its resources for a pre-arranged price [ARECI07].

Example 3. Network operator-equipment supplier product-quality management agreements

To ensure that a product's quality and reliability are priorities of their suppliers, network operators may establish quality-management agreements with their suppliers. Such supplier agreements define a set of quality measures and acceptable product-performance levels. Reliability measurements may include the frequency of outages and system downtime averaged over a year. Quality measurements may cover modulation anomalies such as cross-modulation, interfering harmonics, and quantization impact. Equipment selection depends on the needs of the network, and the criteria included in an agreement must ensure that the necessary quality and other requirements will be met.

A network operator-equipment supplier agreement begins with the network operator specifying the performance requirements for systems and services in the initial purchasing agreements. The same criteria should also be used in evaluating the response of potential suppliers to the quality and reliability requirements. In addition, penalties may be introduced for suppliers whose equipment fails to perform within the agreed range. Methods to assess the quality of equipment and services can take advantage of industry standards, and is described below in the Standards section [TELC05, TL08, NRSC02].

6.3.4 Summary of Agreements

Agreements are mutually accepted terms that define the expectations of two or more parties. The establishment of an agreement is a relatively fast means of controlling the behavior of the entities involved. In the wireless industry, agreements do not require government involvement.

The quality of the wireless experience depends strongly on effective agreements being worked out between network operators and their equipment suppliers. Fortunately, the quality and reliability of network equipment and services can be managed by industry standards developed for that purpose.

6.4 Standards

6.4.1 Background

Standards are a widely accepted means of setting expectations for specific behaviors Other ways of looking at standards in the communications industry include:

- an emphasis on quality: "a set of rules for ensuring quality" [ETSI08]

- an emphasis on process: "a document established by consensus and approved by a recognized body that provides for common and repeated use, rules, guidelines, or characteristics for activities or their results, aimed at the achievement of the optimum degree of order in a given context" [ISO96], or "documents, established by consensus and approved by an accredited standards development organization that provides, for common and repeated use, rules, guidelines or characteristics for activities or their results, aimed at the achievement of the optimum degree of order and consistency in a given context" [IEEE08]

- an emphasis on the voluntary nature regarding following standards: " technical specification approved by a recognized standardization body for repeated or continuous application, with which compliance is not compulsory … "[ETSI08]

Standards enable different networks to interoperate. They allow a large array of wireless handset manufacturers to produce different devices that operate on the same wireless network yet offer the same or similar services. Standards enable different networks to interoperate with the network databases to confirm subscriber billing information. In addition, the logic of the standards development process can add value by helping identify the many possible circumstances that will have to be planned for and dealt with.

Standards are essential elements in helping entities anticipate the behavior of other entities. Compared to regulations, standards are quicker to develop and make available for use. But relative to agreements and policies, they are slower to develop.

6.4.2 Importance of Standards

Without the effective development and implementation of standards, the quality and reliability of services would not be as optimal as they are today. Network elements would fail to interoperate and, at best, user experiences would be inconsistent. Also, having multiple competitors to choose from would be less advantageous to the consumer because of inefficiencies in all of the unique interfaces that have to be managed.

6.4.3   Examples

Additional insight regarding standards can be seen through examples.

Example 1. The wireless standards bodies

A number of standards development organizations play a critical role in regional and international standards writing. These include the following (with examples of the wireless-related areas they address, along with several key standards in Table 6-1):

- Alliance for Industry Solutions (ATIS): *services and systems, network reliability, interconnection with emergency services*

- European Telecommunications Standards Institute (ETSI): *electromagnetic compatibility, emergency communications, reconfigurable radio systems*

- International Telecommunication Union (ITU): *radio regulations, network standards, radio standards, interference standards, electromagnetic compatibility, emergency communications, security, management of radio frequency and satellite orbits*

- International Standards Organization (ISO): *compatibility with medical devices, cabling for wireless access points, radio frequency identification, device testing, performance testing methods*

- Institute of Electrical and Electronics Engineers (IEEE): *wireless local area network (WLAN), quality and reliability, compatibility with hearing aids*

- Internet Engineering Task Force (IETF): *mobility for IP, control and provisioning for wireless access points, mobile ad-hoc networks*

- Telecommunications Industry Association (TIA): *terrestrial mobile multimedia multicast, steel antenna towers, vehicular telematics*

- The 3rd Generation Partnership Project (3GPP): *a collaboration agreement among a number of communications standards bodies; its scope includes 3rd generation mobile systems, enhanced data rates for GSM, and GPRS*

ANSI C63.19:2006	"Methods of Measurement of Compatibility between Wireless Communications Devices and Hearing Aids"—Addresses electromagnetic compatibility between mobile phones and hearing aids.
ETSI EN 300 220-1	Electromagnetic compatibility and radio spectrum matters (ERM); also covers short-range devices (SRD); radio equipment for the 25 MHz-to-1000 MHz frequency range with power levels up to 500 mW; Part 1: Technical characteristics and test methods
IEEE 802.11	"Wireless Local Area Networks"—Standards for wireless LANs that cover modulation techniques; sometimes referred to as Wi-Fi.
ITU-T X.805	"Security architecture for systems providing end-to-end communications"—Provides a common framework for systematically addressing security issues.
TIA-222-G-1	"Structural Standards Abstract for Steel Antenna Towers and Antenna Supporting Structures"—Outlines requirements for structural design and fabrication of structural antennas, support structures, mounts, components, guy assemblies, insulators and foundations.

Table 6-1:   Examples of Standards Applicable to Wireless Networks and Services

Example 2. Describing quality measurements

When an equipment supplier and its many customers have a quality or reliability issue, different words can be used to describe the same situation, or the same words can describe different situations. By setting a common standard, the industry introduces efficiency and intelligence to classifying complex technical situations. In turn, this results in more efficient cost structures and, ultimately, a healthy competitive market and lower costs to consumers subscribing to wireless services.

Several standards help manage the network operator-equipment supplier relationship. Typically, they include definitions and units of measurement. For example, the network reliability performance of a system operating in a live network environment may have reliability measurements expressed in terms of outage event frequency normalized by the number of systems deployed [events/system-year], and also as an expression of downtime [minutes/system-year]. Other quality measurements include the number of software defects, the percentage of defective software patches, and the failure rate of hardware circuit packs.

Example 3. Priority communications

Traffic capacity is one of the fundamental limitations of wireless networks. Limits exist for both the network elements and transport facilities, and for the electromagnetic energy air interface. Because disaster situations are typically accompanied by a greater need for wireless communications, traffic congestion can significantly impair vital emergency communications. To address this, governments have encouraged the development of standards for priority communications. These standards have been developed and implemented voluntarily by the private sector, with government funding as an incentive.

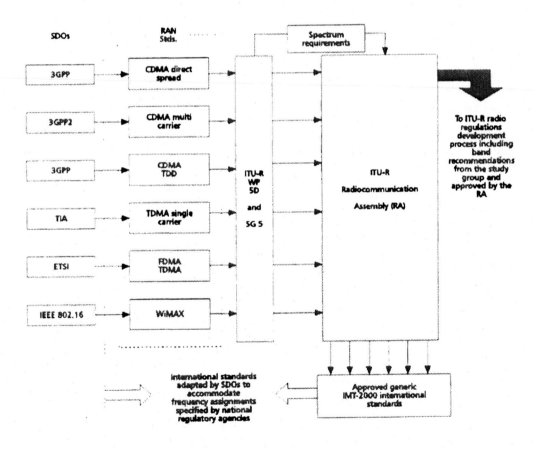

Figure 6-2: Development of Mobile Wireless Telecommunication Standards for IMT-2000

Example 4. Radio interfaces

Figure 6-2 illustrates the process used for a decade in developing specifications for the radio interfaces of International Mobile Telcommunications-2000 (IMT-2000). The organizations at the left of the figure are all international standards development organizations (SDOs). The membership of each includes network operators, service providers, equipment suppliers and other stakeholders in the global mobile wireless communications industry. These SDOs produce radio access network (RAN) standards that are continually being updated. A relatively new member of this group of standards is the mobile WiMAX standard that was developed by the IEEE 802.16 working group. Although not illustrated in Figure 6-2, there is a parallel standardization activity in ITU-T, which develops the network standards for each of the RAN technologies.

To ensure international acceptance of these standards, they are brought into the ITU-R where they are incorporated into an ITU-R Recommendation. After the ITU-R verifies that the submitted standards meet requirements previously specified by the ITU-R, an ITU-R Recommendation is completed and submitted to the Radiocommunication Assembly for final approval. The approved standards, which are often adopted by national standards and regulatory bodies, include the specific bands at which the mobile wireless communication devices can operate in their countries. This interaction between standards development and regulation will be discussed later.

Clearly, spectrum must be made available for these systems. ITU-R working parties create study groups which produce recommendations, and the Radiocommunication Assembly approves spectrum requirements. This is input into the regulatory process described in section 6.6.

6.4.4 Summary

Standards are a widely accepted means of setting expectations for specific behaviors. Key aspects include their ability to promote quality, their development through consensus-based processes, and their voluntary implementation.

Benefits of standards include increased efficiency, interoperability and reliability, and consistency of the user experience. Compared to regulations, standards are quicker to develop and put into use. However, they take more time to develop than agreements and policies.

6.5   Policies

6.5.1   Background

Policies are the guiding principles or plans intended to influence decisions or actions. In the wireless industry, policies help stakeholders prepare for their own future decisions and actions. Policies address issues such as the expectations of the participants dealing with a multi-party outage; procedures for handling non-mandatory government requests for information; the use of environmentally friendly alternatives; corporate posture toward concerns regarding the long-term health effects of using personal wireless devices; and the emergency response to a catastrophe that has a widespread network impact.

Policies can call for government involvement, but do not require it. This is because any entity can establish policies that other entities can expect will influence its behavior. Because the establishment of a policy is within the control of the entity that shapes it, policies are a very fast means of providing guidance for how other entities should anticipate its behavior.

6.5.2 *Importance of Policies*

As mentioned, the term "Policy" can serve as an abbreviation for the longer ASPR phrase. This is because by definition Policy serves as an inclusive umbrella term that includes agreements, standards and regulations. However, Policy has an important distinction: it also includes all the other mechanisms used to anticipate an entity's behavior that none of the other terms capture. For example, industry best practices are neither regulations nor standards, nor do they require an agreement between entities. Yet, industry-consensus best practices play a vital role in promoting the reliability, emergency preparedness, interoperability, and security of wireless networks. Another example is found in the areas of health, safety, and the environment. While some regions of the world have regulations to control behaviors in these areas, their approaches will fall short in some situations of being complete regarding what is appropriate. Many companies have policies that other entities, when interacting with them, can depend on, such as being able to expect that a company's facilities will be designed and operated with health and safety considerations in mind.

6.5.3 *Examples*

Additional insights into policies can be gained through examples.

Example 1. Industry-consensus best practices
The wireless industry has developed hundreds of consensus best practices. Companies that support such development are demonstrating a policy of expert information sharing with their industry peers on critical aspects of wireless networks in order to promote their reliability, emergency preparedness, and security. Because the implementation of a given set of best practices is not applicable in all situations, their implementation is voluntary. They're not applicable in all situations because of such things as differences in some emerging technologies, the use of an alternative or otherwise unique configuration, and the need for site-specific rules for special considerations.

Industry-consensus wireless-network best practices can be accessed on the Internet at

> http://www.bell-labs.com/USA/NRICbestpractices/, and
>
> http://www.bell-labs.com/EUROPE/bestpractices/ .

Table 6-2 provides several examples of industry best practices.

7-7-0454	Network Operators and Service Providers should consider establishing technical and managerial escalation policies and procedures based on the service impact, restoration progress, and duration of the issue.
7-7-0495	Network Operators and Property Managers should consider pre-arranging contact information and access to restoral information with local power companies.
7-7-0578	Network Operators, Service Providers and Public Safety Agencies should actively engage in public education efforts aimed at informing the public of the capabilities and proper use of 911.
7-7-0472	Network Operators and Equipment Suppliers should consider connector choices and color coding to prevent inappropriate combinations of RF cables.

Table 6-2: Best Practices for Wireless Networks

Example 2. Crisis support

Many consumers buy a wireless device and service to enhance their personal security. Realizing how important mobile phones are, some wireless service providers have set policies that continue to allow people to depend on them during a crisis.

For example, service providers will set up phone banks for the victims of a catastrophe so they make free phone calls. Alternatively, providers will allow subscribers an automatic 2- to 3-month delay in paying their bills due to the hardship. In still another example, some wireless service providers will provide public safety workers—and even insurance adjusters—with free mobile phones to support their critical work.

Perhaps one of the most impressive examples of cooperative response a during major catastrophe is when the industry comes together to support volunteers from among its ranks in the use of advanced wireless technology to enhance traditional search-and-rescue efforts. In such circumstances, a wireless device may be a victim's only hope for survival. The Wireless Emergency Response Team (WERT) headquartered in the Lehigh Valley, Pennsylvania, USA, is a non-profit organization that coordinates such volunteers.

WERT's vision is to "connect the best minds and resources of the wireless industry to the most vital needs of its subscribers" [WERT07]. With the combined resources of expert volunteers and portable network equipment, WERT has been able to simulate wireless networks that have been destroyed or greatly impaired. In addition, network operations experts have been able to analyze wireless emergency traffic during crises to identify emergency calls that may be receiving insufficient attention.

Example 3. Emergency back-up power

Communications are vital during a crisis. One of the most common crises encountered around the world is the loss of commercial electric power. Personal mobile devices depend on a small battery that can typically last up to several days or provide several hours of talk time. However, during a power outage, the mobile device can only make a call or send a text message if a nearby cell site and its backhaul network have emergency backup power. Recognizing the vital importance of communications during a crisis, many network operators have voluntarily deployed emergency backup power capabilities at cell sites. Typically, these capabilities consist of batteries or generators, or a combination of the two. Some network operators also reach out to the local commercial electricity provider to improve their coordination during the restoration activities after a disaster.

It has been noted that the minimum backup power appropriate for a given cell site depends on the circumstances of that particular site [CQR04]. Underscoring this, industry-consensus expert guidance was provided to the U.S. Federal Communications Commission (FCC) and the private sector, as part of an official U.S. Federal Advisory Committee Act (FACA), as follows:

> Network operators should provide backup power (e.g., some combination of batteries, generator, and fuel cells) at cell sites and remote equipment locations, consistent with the site-specific constraints, criticality of the site, the expected load and reliability of primary power [NRIC05*].

An advantage of voluntary policies is they allow limited resources to be invested in technologies that can optimize the benefits provided to end users. For example, while a strategy to have a minimum number of, say, x hours of backup battery power at all cell sites is simple to understand, it does not address needs beyond x hours. On the other hand, some wireless network operators have found that, at times, they can better meet the needs of their subscribers in a crisis by providing virtually unlimited generator-based power at a small number of interspersed sites. At these sites the reception gain and transmission power are temporarily increased to extend into adjacent geographic areas normally covered by other cell sites, which because of the crisis, may have no power. This example highlights the value of policies whereby experts based have developed solutions that optimize the benefits to consumers.

6.5.4 Summary

In the wireless industry, policies help stakeholders prepare for their own future decisions and actions. Policies do not require government involvement, as entities can themselves establish policies that other entities can expect will influence their behavior. However, there are government policies that impact wireless communications. Because establishing a policy is within an organization's own control, setting policies is a very fast way to provide guidance for how other entities should anticipate its behavior.

One of the most popular forms of a wireless industry policy is voluntary, industry-consensus best practices. Distinct from regulations and standards, best practices play a vital role in promoting the reliability, emergency preparedness, interoperability, and security of wireless networks. However, best practices are not applicable in all situations for several reasons, including differences in emerging technologies, the presence of alternative or otherwise unique configurations, and the need for site-specific rules for special considerations.

While some regions in the world have regulations governing behavior regarding health and safety, such approaches will fall short of what is appropriate in all situations. Accordingly, many companies have policies that other entities interacting with them can depend on, such as being able to expect that a company's facilities will be designed and operated with appropriate health and safety considerations. "Another important organization is The Association of Public-Safety Communications Officials–International (APCO), which supports the wireless communication services needs of police, fire, EMS, and other public safety agencies worldwide." In addition, APCO plays a critical role in the communications standards used by different public safety agencies throughout the world—for example, the TETRA (TErrestrial Trunked RAdio) standard in Europe and the P25 (Project 25) standard in the U.S. are digital wireless communication standards developed by the organization [APCO08].

6.6   Regulations

6.6.1 Background

Regulations are rules established by government authorities; they are not optional. When a regulation is not obeyed, the violator can face negative consequences, such as having to submit to a formal hearing, pay a fine, or have its license suspended or revoked. Regulations are necessary to ensure coordination among entities in some circumstances.

Regulations are characteristically slow to develop relative to the pace of the wireless industry's growth **and introduction of new technology. This is due in part to the inclusion of checks and balances appropriate to the development of public policy. Once established, regulations are also hard to change. Another characteristic is that regulatory policy struggles with areas of rapidly developing technology. This is due in part to the level of expertise and experience required to effectively deal with such areas; most regulators are primarily trained as lawyers, economists, or in other disciplines. Seldom are they engineers with technical expertise in the emerging technologies they are regulating. It is also difficult to develop regulations that can predict the impact of new technologies.**

Finally, regulations use the force of the law to control the behavior. This is constraining and undesirable from the perspective of wireless industry stakeholders, who for the most part, are continually upgrading the capabilities they offer. Because of these inherent attributes of the regulatory process, regulations are often seen as the last resort for controlling the behavior of industry entities. "Regulatory agencies sometimes assign unlicensed bands such as those in the 2.4 GHz as well as 5 GHz (WiMAX) bands. Due to basic guidelines but lack of regulation, these bands can become so polluted that in many metropolitan areas the noise floor increases sharply, making it difficult for many designers to use these bands effectively [CONV07, WNDL06]."

The wireless industry typically must submit to several levels of government authority, such the federal and regional authorities and the European Union, its member states, and their local municipalities.

Because of the potential impact that regulations can have on their interests, private sector companies have organized themselves to work together to ensure their points of view are heard during the development of public policy. CTIA, the international association for the wireless telecommunications industry, is an example of such an organization [CTIA08]. An example of an industry association that focuses on infrastructure (towers, cell sites, etc.) is PCIA, the wireless infrastructure association [PCIA08].

6.6.2 Importance of Regulations

Regulations are needed when other means of coordinating behavior among entities are ineffective. Reasons can include gridlock caused by competitive forces, a lack of incentives to act, or the need for a framework that establishes a level competitive playing field. Another reason for relying on regulations is if the risks of unacceptable behavior can adversely affect public safety or other important social or nation-state security concerns.

6.6.3 Examples

Additional insights regarding regulations can be seen through examples.

Example 1. Wireless frequency spectrum management
The wireless transmission portion of a call path relies on invisible electromagnetic energy within a set frequency range. Because the frequency spectrum is limited, managing its use is of great concern. Misuse can cause interference with other wireless carriers and electronics. Because of such unacceptable consequences, government authorities have stepped in and regulated aspects of the spectrum. For example they have:

- Issued licenses to users of specific frequency bands [FCC08].

- Addressed cross border situations, where the transmission and reception of the RF signal crosses into another country.

- Prohibited frequency jammers.

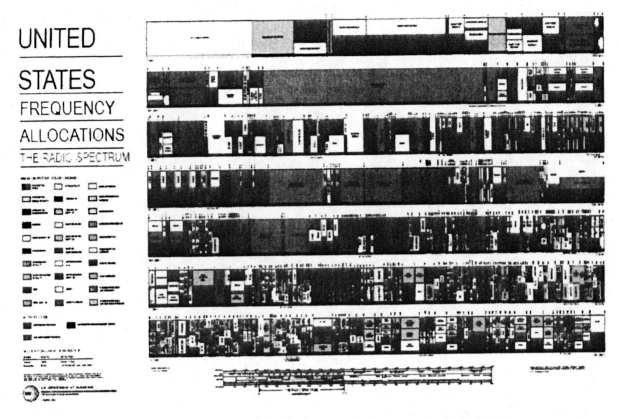

Figure 6-3: Frequency Allocations of the Radio Spectrum in the United States

Figure 6-3 shows the frequency allocations for the entire radio spectrum managed in the United States by the National Telecommunications and Information Administration (NTIA). One of purpose of showing this figure is to indicate how complex the management of frequency spectrum is [FCC08*]; other countries have similar databases and tools.

One familiar example of the negative impact of not managing the spectrum correctly is the interference heard on GSM telephones, speakers, car radios, home computer speakers, etc.

Figure 6-4 depicts an example of an international and national regulatory process. Specifically, it is for mobile wireless telecommunication services and is a continuation of Example 4, *Radio Interfaces*, described earlier in Standards section 6.4.3. One should keep in mind that these processes are both iterative and interactive. Approved ITU-R Recommendations on spectrum requirements are part of the input to the World Radiocommunication Conference (WRC), including the Conference Preparatory Meeting. A primary product from the WRC is the Radio Regulations document, which is a treaty-level document since the ITU is chartered by the United Nations.

One of the most critical parts of the Radio Regulations is Article 5 of Volume 1. This is the Table of Frequency Allocations, which allocates specific bands to specific services globally. New requirements for land mobile services, for example, are discussed during the WRC and changes are made to the Table of

Frequency Allocations by international agreement. Because of the competing demands for spectrum, this is a very difficult process.

The ITU-R Radio Regulations provide an input to the regional and national regulatory bodies that make decisions on frequency assignments within the allocated bands. This process is indicated at the bottom of Figure 6-4.

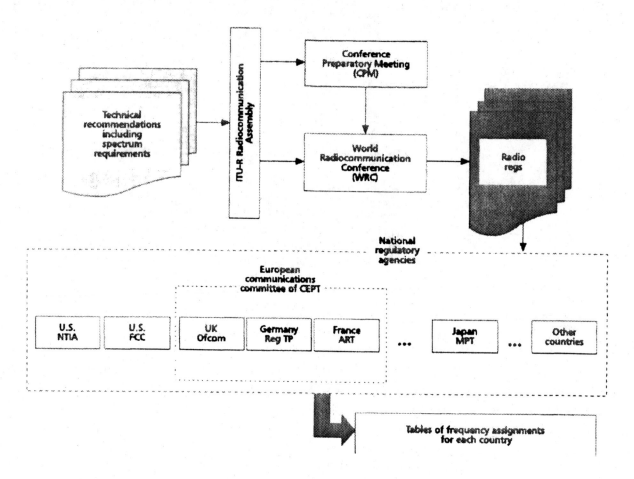

Figure 6-4: Block Diagram of the Process for Developing International and National Mobile Wireless Telecommunication Regulations

Example 2. Promoting investment in technology

One of the benefits of regulating the spectrum appreciated by the commercial sector is the certainty it brings. When it is known that certain frequencies will be allocated to wireless services for a long time, companies can move forward with confidence that the enormous investment needed to develop equipment and build networks has a chance of yielding a financial return.

Similarly, in the U.S., the FCC has taken steps to "remove regulatory barriers and facilitate the development of secondary markets in spectrum usage rights among the Wireless Radio Services" [FCC03]. The objective of these regulatory changes is to facilitate broader access to spectrum resources using spectrum-leasing arrangements.

European Commission EC No 717/2007	**Roaming on public mobile telephone networks** Limits roaming charges to consumers using mobile devices outside their home region.
U.S. Federal Communications Commission FCC 47 CFR 4	**Disruptions to communications** Requires wireless service providers to report outages to the FCC that meet specified criteria.
Denmark National IT and Telecom Agency Consolidated Act No. 663	**Joint utilization of masts for radio communications purposes** Requires owners of masts for radio communications to allow other parties to jointly use their masts.
ITU-R Radio Regulations	**Frequency allocations** Provides an input to the regional and national regulatory bodies that make decisions on frequency assignments within the allocated bands.

Table 6-3: Examples of Regulations Applicable to Wireless Networks and Services

6.6.4 *Summary*

Regulations are needed when the risks of an entity behaving in an unacceptable way can lead to unacceptable consequences for public safety or other important social or nation-state security concerns. The wireless industry typically must submit to several levels of government authority.

Relative to the pace of the wireless industry's growth and introduction of new technology, regulations are characteristically slow to develop. Because of the inherent attributes of the regulatory process, regulations are often seen as the last resort for controlling the behavior of entities.

Violators of a regulation can face negative consequences. They could be required, for example, to submit to a formal hearing, pay a fine, or have their license suspended or revoked.

6.7   References

[ARECI07] *Availability and Robustness of Electronic Communications Infrastructures (ARECI) Study Final Report*, Formal Mutual Aid Agreements, Recommendation 2, pp 17, 56, 99; March 2007, ec.europa.eu/information_society/newsroom/cf/itemdetail.cfm?item_id=3334.

[ATIS07] "A term used to refer to the complete set of inter-entity arrangements that are necessary for communications services; these arrangements include national and international standards; federal, state and local regulations or other legal arrangements; or any other agreement between entities—including industry cooperation and agreements and other interfaces between entities. ASPR is one of the elements of the Eight Ingredient Framework; in this framework, it is sometimes abbreviated as "Policy." The term ASPR is used in several of the many places the Eight Ingredient Framework, including, most recently, the President's National Security Telecommunications Advisory Committee (NSTAC) "Next Generation Networks Task Force Final Report," March 2006. ATIS Telecom Glossary 2007, www.atis.org.

[APCO08] http://www.apco911.org

[CONV07] http://www.convergedigest.com/bp-bbw/bp1.asp?ID=391

[CQR04] Slide 5 of Survey Questions, IEEE Communications Society Technical Committee on Communications Quality & Reliability (CQR), Proceedings of the Emergency Power Conference, Washington, D.C., November 2004; "58% of Subject Matter Experts Agreed That the Minimum Number of Hours of Backup Power Depends on a Number Of Factors."

[CQR07] *IEEE Communications Society Technical Committee on Communications Quality & Reliability (CQR)*, Proceedings of the 2001 CQR International Workshop, May 2001.

[CTIA08] www.ctia.org

[FCC2008] http://www.fcc.gov/spectrum/

[FCC08*] The FCC maintains tools for managing this information, see http://www.fcc.gov/oet/info/database/

[FCC03] http://wireless.fcc.gov/licensing/index.htm?job=secondary_markets

[IEEE08] IEEE 100, The Authoritative Dictionary of IEEE Standards Terms, Seventh Edition.

[ISO96] ISO/IEC Guide 2:1996, definition 3.2

[NRSC02] The Alliance for Telecommunications Industry Solutions (ATIS) Network Reliability Steering Committee (NRSC), *The NRSC 2002 Annual Report.*

[NRIC03] The Federal Communications Commission (FCC) Network Reliability and Interoperability Council *NRIC VI Homeland Security Physical Security Focus Group Final Report*, Issue 3, December 2003.

[NRIC05] *NRIC VII Wireless Network Reliability Focus Group Final Report*, Issue 3, October 2005, NRIC VII Public Data Network Reliability Focus Group, Issue 3, October 2005.

[NRIC05*] Best Practice Number 7-7-0492, p. 52 *NRIC VII Wireless Network Reliability Focus Group Final Report*, Issue 3, October 2005.

[NSTAC06] The President's National Security Telecommunications Advisory Committee (NSTAC) *Next Generation Networks Task Force Report*, 2006.

[PCIA08] www.pcia.org

[Rau06] K.F. Rauscher, R.E. Krock, J.P. Runyon, *Eight Ingredients of Communications Infrastructure: A Systematic and Comprehensive Framework for Enhancing Network Reliability and Security*, Bell Labs Technical Journal, Vol. 11, No. 3, Wiley InterScience, 2006.

[TELC05] GR 1929-CORE, *Reliability and Quality Measurements for Telecommunications Systems (RQMS-Wireless)*, Telcordia.

[TL08] www.tl9000.org.

[WERT2007] Wireless Emergency Response Team website: www.wert-help.org.

[WNDL06] http://www.wirelessnetdesignline.com/howto/60401206

John Wiley and Sons, 2003.

Chapter 7

Fundamental Knowledge

7.1 Introduction

This chapter outlines the fundamental subjects a wireless communication engineer should know. It is the foundation on which to build advanced knowledge and skills. Typically, the subjects are covered by most engineering schools in a basic 3- to 5-year educational program.

The basic knowledge needed to perform satisfactorily is very broad. It would take many pages to cover everything. Therefore, only high-level descriptions are presented here, no detailed mathematics, investigation, or analysis. In many cases, space limitations preclude doing anything more than listing under each topic the concepts with which a wireless communication engineer should be familiar. However, references are included to several of the many textbooks in which the subjects are well covered. It is up to the individual engineer to acquire the fundamental knowledge needed to function in the profession, and to understand the topics described in the chapters of this book.

Certainly, the wireless communication engineer must be knowledgeable in electrical and communications engineering, and must have a general understanding of economics and project management. Basic knowledge of computers, microprocessors, algorithms, programming, and basic software are assumed. These tools are very important in the daily duties of any engineer.

7.2 Contents

This chapter covers the following topics:

Electrical Engineering Basics for Wireless Communications

Signal Processing and Communication Systems

RF Engineering

Instruments and Measurements

Communication Networks

Other Communication Systems

General Engineering Management and Economics

7.3   Electrical Engineering Basics for Wireless Communications

7.3.1 Fundamentals of Electronics [Mill87] [Tho06]

Fundamentals are mostly covered in undergraduate electrical/electronic engineering degree programs. Not all engineers may be directly involved in the design and development of electronic modules for wireless communications once they graduate. However, all engineers must know the fundamentals. Analog and digital circuits, at high or low frequencies, form the backbone of the hardware used for wireless communications, either in the network or on the customer's premises. The following list of basic topics includes the concepts a wireless communication engineer should know and understand.

Linear Circuits [Tho06]: Includes voltage and current, resistors (R), capacitors (C), inductors (L), voltage and current sources, Ohm's law, power and energy, Kirchhoff's current and voltage laws, node-voltage and mesh-current circuit analysis, Thevenin and Norton equivalent-circuit representations, theory of superposition for linear circuit analysis, and equivalence and reciprocity theorems.

Inductors and Capacitors [Tho06]: Concepts include the analysis of RL, RC, and RLC circuits, natural and step response, series and parallel RLC circuits, resonant frequency and Q factor, higher order RLC circuits, the sinusoidal source, sinusoidal response, phasor analysis, frequency response, basic filters, and frequency-selective circuits.

Electric Power [Tho06]: Includes such concepts as instantaneous, average, complex, real and reactive powers, root mean square (rms) value, power factor, and maximum power transfer conditions.

Magnetically Coupled Circuits and Transformers [Tho06]: Concepts include electric and magnetic fields, flux density, self and mutual inductance, and current and voltage transformers.

Amplifier Circuits [Mill87]: Includes the PN junction, BJT and FET transistors, basic transistor circuits, biasing, current- and voltage-mode amplifiers, positive and negative feedback, gain-bandwidth product, input/output impedance, frequency response, 3-dB bandwidth, current and voltage gain, and cascading amplifiers.

Operational Amplifiers [Mill87]: Included among the concepts are basic characteristics, open/closed loop gain, inverting/non-inverting amplifiers, the OPAMP as a basic circuit-building block, unity-gain buffers, slew rate, and other similar specifications.

Boolean Algebra [Lev79]: Concepts include fundamental theorems of Boolean algebra and Boolean functions (truth tables and canonical forms, representation forms, and main properties).

Logic Circuits [Mill87]: Concepts addressed under this topic include discrete and integrated logic gates such as AND, OR, NAND, and NOR, and their interconnections, design of simple logic using transistors, binary logic, and truth tables.

Although, this list focuses specifically on the basic hardware fundamentals, most modern electronics depend on software, as well. The hardware may include one or more microprocessors, field programmable gate arrays (FPGAs), analog or digital signal processing chips or application specific integrated circuits (ASICs), peripherals, and input/output circuitry. Programmed with software, the microprocessors and FPGA chips are interfaced to other hardware. The wireless communication engineer may need to work with such electronic boards, perhaps to develop and test new algorithms or simply to troubleshoot them.

7.3.2 Electronic Power Supply Design

A power supply refers to a system that supplies electrical energy to an output load or group of loads at certain specifications (output voltage, current, ripple, etc.). In general, such supplies convert one form of electrical power to another form and voltage level. For electronic devices this typically involves converting 120- or 240-volt AC supplied by a utility (see electricity generation) to a well-regulated lower-voltage DC. Topics to be aware of include:

- Power supply definitions
- Power adaptors
- Step-down transformers
- Rectifiers and inverters
- Linear regulators
- Switched-mode power supplies
- Voltage stability
- Ripple

- Surge protection
- Current/voltage regulation
- Continuous/pulse-mode supplies
- Uninterruptible power supply (UPS)
- Short-circuit protection
- Overpower (overload) protection
- Over-voltage protection.

7.3.3 Basic Printed Circuit Board Design Considerations [Kha05]

A printed circuit board or PCB mechanically supports and electrically connects electronic components using conductive pathways, or traces, etched from copper sheets laminated onto a nonconductive substrate. The PCB plays an important role in miniaturizing portable devices. The wireless engineer should know how a bare PCB is prepared, including its artwork (for the printed circuit design), printed circuit assembly, the soldering process, and the testing and quality control of the final product. Some of the key topics related to PCB manufacturing are presented in Table 7-1.

PCB Types	Laminate Materials	Integrated Circuit (IC) Packaging Techniques
Single-sided/double-sided Surface mount or through hole Single- or multi-layer Flexible or rigid	FR-4 (the most common PCB material), FR-2, polyimide, GETEK, BT-Epoxy, cyanate ester, Pyralux for flexible printed circuits, PTFE (polytetrafluoroethylene), Rogers Bendflex, conductive ink	Dual in-line package (DIP), pin grid array (PGA), leadless chip carrier (LCC), surface mount with either gull-wing or j-leads, small-outline integrated circuit (SOIC), plastic leaded/leadless chip carrier (PLCC), plastic quad flat pack (PQFP), thin small-outline package (TSOP), land grid array, flip chip, ball grid array (BGA), multi-chip module, System in Package (SiP), hybrid integrated circuit (HIC), chip-on-board (COB)

Table 7-1: Key Topics in PCB Manufacturing

Wireless communication engineers should have some familiarity with the following subject areas and the topics listed under each of them.

<u>Bare Board Testing</u>: Topics include testing before populating the PCB and testing methods—bed of nails, flying-probe, and flying-grid testing.

<u>Computer-aided Layout Preparation:</u> Includes Gerber files, netlists, numerical control drill file and assembly drawings, moisture absorption, thermal-expansion issues.

<u>Electromagnetic Interference Compatibility/Electromagnetic Compatibility</u>: EMI and EMC issues are very important in preparing a PCB layout at microwave frequencies. Topics include the transmission line characteristics of traces, EMI characteristics of strip and micro-strip lines, RF loop current issues, EMI filtering using inductors and bypass capacitors, FCC Part 15 (covering intentional and unintentional EM transmitters and receivers), and other regulations.

<u>Grounding and Electrostatics</u>: Includes grounding and isolation of analog/digital and mixed grounds, providing ground planes, potentially harmful electrostatic discharge (ESD) issues, and shielding against ESD and surges.

<u>Assembly</u>: Topics include various mounting techniques such as through-hole and surface mounting (SMT) for single/double-sided assembly, and various soldering techniques, including hand, wave and reflow soldering.

<u>Test after Assembly</u>: Testing is essential for assessing the functioning of PCBs. Methods include in-circuit and boundary-scan tests, and test standards from the Joint Test Action Group (JTAG).

<u>Standards Bodies</u>: Among the many standards bodies are the Association Connecting Electronics Industries, known as IPC, and the Joint Electron Device Engineering Council (JEDEC).

<u>Specifications</u>: Relevant specifications include MIL-PRF-50884, MIL-PRF-31032, MIL-PRF-55110, UL 796, and various safety standards for PCBs.

7.4   Signal Processing and Communication Systems

This section summarizes the background needed for a solid knowledge of communications, including general techniques for wired and wireless communications with special emphasis on wireless. Important elements are the basics of analog and digital modulation, techniques for analyzing channel characteristics, modeling and treating noise, and EMC issues.

7.4.1 Basic Signal Processing [Pro06]

Topics and concepts with which a wireless communication engineer should be familiar include the following.

<u>Multiple Discrete Random Variables</u>: Includes transforms, prediction, covariance, and correlation.

<u>Jointly Distributed Random Variables:</u> Included here are definitions, calculations, independent RVs, and sums.

<u>Basic Theorems</u>: Include the weak law of large numbers, the central limit theorem, and the strong law of large numbers.

<u>Descriptive Numerical Statistics</u>: Concepts include mean, median, standard deviation, variance, range, and inter-quartile range.

Properties of Signal and Noise: These include deterministic and random signals, continuous and discrete time signals, power and energy signals, periodic signals, signals operation like scaling and shifting, mean value and mean square value (power) of a signal, root-mean-square (RMS) value, the impulse (Delta) function and unit-step function, (non)stationary signals, Ergodicity, white noise, noise-equivalent bandwidth, noise probability density function, and Gaussian noise.

Signal Transmission with Noise: Included are additive noise and signal-to-noise ratio (SNR) and SNR in various carrier-modulated and baseband systems.

Signal Relationships: Includes auto-correlation, cross-correlation, orthogonal signals, convolution and de-convolution operations, convolution with the impulse function, and the shifting property.

Sampling: Concepts include sampling frequency and theorem, Nyquist rate, quantization, digital to analog conversion (DAC), analog to digital conversion (ADC), aliasing, and sample-rate conversion (decimation, up/down sampling interpolation).

Fourier Series: Included are trigonometric Fourier series and complex exponential Fourier series, properties of Fourier series, symmetric spectra for real signals, and Parseval's relation.

Fourier Transform: Concepts include time- and frequency-domain concepts, definition and properties of the Fourier transform; Parseval's theorem and energy spectral density, band-limited signal and noise, and discrete and fast Fourier transform (DFT/FFT).

Other Transforms and Filters: These include Z-transform, digital filters (finite impulse response–FIR, infinite impulse response–IIR, and lattice), high-pass filter (HPF), low-pass filter (LPF), band-pass filter (BPF), all-pass Filter (APF), and pole-zero plots.

7.4.2 Communications and Information Theory [Pro07]

A wireless communication engineer should have at least a basic understanding of the following topics and concepts.

Basics: These include bit time, baud time, symbol time, inter-symbol interference (ISI), crest factor, peak-to-average power ratio (PAPR), complex envelope I & Q waveforms, Shannon theorem, coherent detection, con-coherent detection, geometrical representation, differential encoding and decoding, differential detection, constellation diagram, eye diagram, trellis diagram, signal-space representation, bit error rate (BER), symbol error rate (SER), Q function, antipodal signaling, constant envelope, probability upper bound, signal-to-noise ratio per bit (Eb/No), spectral analysis, occupied BW, effect of transmit non-linearity on modulation schemes, and spectral re-growth.

Characterization of Communication Signals and Systems: Concepts include signal space representations, spectral characteristics of signals, bandpass signals and systems, complex baseband representation.

Analog Signal Transmission and Reception: Included are introduction to modulation, amplitude modulation (AM), angle modulation, and effect of noise on analog communication systems (effect of noise on linear-modulation systems, carrier-phase estimation, effect of noise on angle modulation, effects of transmission losses and noise in analog communication systems).

Digital Transmission Through the Additive White Gaussian Noise (AWGN) Channel: Includes geometric representation of signal waveforms, pulse-amplitude modulation, two-dimensional signal waveforms, multidimensional signal waveforms, optimum receiver for digitally modulated signals in additive white Gaussian noise, probability of error for signal detection in additive white Gaussian noise, performance analysis for wire-line and radio-communication channels, and symbol synchronization.

Digital Transmission Through Band-limited AWGN Channels: Concepts include the power spectrum of digitally modulated signals, signal design for band-limited channels, probability of error in detection of digital pulse amplitude modulation (PAM), digitally modulated signals with memory, system design in the presence of channel distortion, multicarrier modulation, and orthogonal frequency division multiplexing (OFDM).

Frequency and Phase Modulation: Included are the relationship between FM and PM, instantaneous and peak frequency deviation, Carson's rule, narrowband and wideband FM, modulation index, Bessel functions, capture effect, phase nonlinearities, AM-to-PM conversion, and pre-emphasis and de-emphasis.

Digital amplitude modulation	On/off keying (OOK), amplitude shift keying (ASK), pulse amplitude modulation (PAM), quadrature amplitude shift keying (QASK), quadrature amplitude modulation (QAM), M-ary QAM
Digital frequency modulation	Frequency shift keying (FSK), binary frequency shift keying (BFSK), M-ary FSK, fast frequency shift keying (FFSK)
Digital phase modulation	Phase shift keying (PSK), binary phase shift keying (BPSK), differential phase shift keying (DPSK), Quaternary phase shift keying (QPSK), offset QPSK (OQPSK), sinusoidal OQPSK, M-ary PSK
Continuous phase modulation	Minimum shift keying (MSK), Gaussian minimum shift keying (GMSK)
Pulse shaping	Raised cosine filtering, matched filter, correlation, union-bound approximations, pulse shaping, bandwidth efficiency (b/s/Hz)

Table 7-2: Schemes for Digital Modulation and Pulse Shaping

7.4.2.1 Coding and Error Correction

Source coding: Source coding compresses data to be transmitted, which increases the entropy in each symbol. Entropy is each symbol's actual information content.

Topics include mathematical models for information sources, measures of information, entropy and mutual information, asymptotic equipartition property, entropy rate, source-coding theorem, source-coding algorithms, coding for discrete sources, coding for analog sources, optimum quantization, rate-distortion theory, and waveform coding.

Channel capacity and coding: Channel coding reduces information rate through the channel and increases reliability. This is achieved by adding redundancy to the information symbol vector, resulting in a longer coded vector of symbols distinguishable at the output of the channel. Channel coding can be performed with block or convolutional codes.

Topics include channel models and channel capacity, bounds on communication.

Types of codes:

    Linear block codes: The information sequence is divided into blocks of length k. Each block is mapped into channel inputs of length n $(n>k)$. The mapping is independent of previous blocks—that is, there is no memory from one block to another. Examples include Hamming code, BCH code, Reed-Solomon code, Reed-Muller code, Binary Golay code, and low-density parity-check codes.

<u>Convolutional codes</u>: Each block of k bits is mapped into a block of n bits but these n bits are not only determined by the present k information bits but also by the previous information bits. This dependence can be captured by a finite-state machine. Topics include Viterbi coding, punctured convolutional codes, trellis diagrams, and turbo coding.

<u>Error detection</u>: This includes simply detecting the presence of error, as well as parity check and cyclic redundancy check (CRC).

<u>Error correction</u>: This involves an additional ability to reconstruct the original error-free data

<u>Automatic repeat-request (ARQ)</u>: The receiver requests retransmission in case of error. Often, retransmission will begin if the transmitter does not receive an acknowledgement (ACK) of correctly received data in a reasonable time.

<u>Forward error correction (FEC)</u>: The transmitter encodes data with an error-correcting code and sends the coded message. The receiver decodes what it receives into the "most likely" data. (FEC can be combined with ARQ.)

7.5   RF Engineering

7.5.1   *Basic Electromagnetic Waves and Transmission Lines [Kra99]*

The wireless communication engineer clearly needs to have a thorough understanding of electromagnetic waves and transmission lines, including the following key topics and concepts:

<u>Electromagnetic Waves</u>: Includes dominant frequency bands, Maxwell's equations, electromagnetic wave equations, scalar and vector potentials, propagation properties, power density, and polarization (horizontal and vertical).

<u>Types of Physical Noise</u>: Includes thermal noise (Johnson noise or Nyquist noise), shot noise, pink noise (flicker noise) or 1/f noise, Brownian noise, burst noise, phase noise, click noise, noise temperature, and noise figure.

<u>Radio Wave Propagation</u>: Concepts included are wave basics, the Friis formula, free space path loss, reflection, refraction, and diffraction; very-low-frequency (VLF), low-frequency (LF), and medium-frequency (MF) propagation; HF propagation, VHF and UHF propagation, microwave propagation; Snell's law, reflection and transmission coefficient, critical angle, Fresnel's equations, normal and vertical polarization, oblique incidence, Brewster's angle, and reflection from and transmission through planar slabs.

<u>Transmission Lines</u>: Included are distributed equivalent circuit, transmission line (telegrapher's) equations, field analysis of transmission lines and per-unit-length parameters, solutions to transmission line equations, characteristic impedance, transmission line terminations, reflection and transmission coefficients, standing wave ratio, impedance matching, admittance, Smith chart, calculations with the Smith chart, and voltage standing wave ratio (VSWR).

<u>Types of Transmission Lines</u>: Includes coaxial cables, metallic waveguides, modes in waveguides, planar transmission lines, microstrips, striplines, coplanar waveguides, dielectric waveguides, coupled microstrips and striplines, coupled transmission lines in printed circuit boards, and integrated circuits.

<u>Parameters Related to Antennas</u>: Concepts include electromagnetic radiation pattern, half-power beam width, directivity, radiation resistance and efficiency, power gain, bandwidth, effective area, reflection

coefficient, input impedance, bandwidth, efficiency, polarization, front-to-back ratio, antenna factor, mutual impedance, transmitting and receiving antennas, reciprocity, receiving antenna aperture, wideband antennas, and polarization matching.

7.5.2 Basic EMI, EMC, and Interference [Pau06]

An understanding of the following topics and concepts related to electromagnetic interference (EMI), electromagnetic compatibility (EMC), and other forms of interference is important in assuring the proper operation of a wireless network.

Basics of EMI and EMC: Concepts include sources of EMI, feedback mechanisms, protection schemes, basic techniques for EMI analysis and modeling, interconnections and wiring for EMI reduction, ground design, EMI filters, digital-circuit, common-mode and differential-mode emissions, and common-mode filters.

Electromagnetic Shielding: Includes basic theory, screening techniques, shielding materials and geometric structures, shielding for EMI protection, and cable shielding and termination.

Intrinsic Noise in Circuit Components: Concepts include noise generated by devices and active components, noise in digital circuits, noise suppression techniques, and transient suppression.

Radiation in Digital and Microwave Circuits: Included are electric and magnetic field coupling, transmission line effects and terminations, radiation by gaps, holes, and loops in printed circuit boards, and practical techniques to reduce coupling and radiation in electronic systems.

Grounding: Concepts include EMC design of ground planes, ground grids, ground inductance, substrate coupling, closing current loops, AC power grounds and power-supply isolation, ground loops, equipment and enclosure grounding, mixed-signal printed circuit boards, return current paths, PCB partitioning and split ground planes, and bridges.

Electrostatic Discharge (ESD): Includes electrostatic fields, protection from ESD, metal enclosures, non-metallic enclosures, and ESD immunity.

EMI/EMC Measurements: Included are current and field probes, transmitted and radiated noise, electric field, magnetic field, power density, electromagnetic susceptibility, shielding effectiveness, absorption and reflection loss, and open-field and anechoic-chamber tests

EMC Legislation and Regulations: Important references include FCC Part 15 and CISPR (The International Special Committee on Radio Interference) regulations.

7.5.3 RF Circuitry, Components and Design [Whi03]

Important RF circuit-design concepts include building blocks in RF systems, RF filter design, matching networks, active and passive device characteristics, and modeling.

Basic Circuits: Concepts include lumped elements, RLC circuit elements, design of lumped elements, lumped element modeling, fabrication, and applications.

Inductors: Includes basic definitions, inductor models, coupling between inductors, electrical representations, printed inductors on substrates, and wirewound and bond wire inductors.

<u>Capacitors</u>: Included are capacitor parameters, chip capacitors, parallel-plate capacitors, voltage and current ratings, and monolithic capacitors.

<u>Resistors</u>: Concepts include types of RF resistors and high-frequency resistor models.

<u>RF Transformers</u>: This includes basic theory, wire-wrapped transformers, transmission-line transformers, ferrite transformers, parallel-conductor-winding transformers on silicon substrates, and spiral transformers on GaAs (gallium arsenide) substrates.

<u>Basic RF Building Blocks</u>: Included here are basics of RF transceivers; impedance matching and impedance-matching circuits, low-noise amplifiers, mixers, filters, oscillators and power amplifiers, and nonreciprocal components.

<u>Parameters of Multiport RF and Microwave Circuits</u>: Concepts include S-parameters, Z- and Y-parameters, ABCD parameters, and conversion formulas.

<u>Active RF Components and Modeling</u>: Included are RF diodes and transistors, modeling, measurement and characterization of active devices; the substrate effect, power loss in substrates, cutoff frequency (at which current gain is unity), 1/f noise and noise figure (NF), frequency synthesizers, and basic design of RF circuits.

<u>RF and Microwave Resonators</u>: This includes lumped RLC, transmission-line and dielectric resonators, cavities, and resonator coupling and excitation.

<u>Power Dividers and Directional Couplers</u>: Concepts include properties of three- and four-port circuits, t-junction power and Wilkinson power dividers, waveguide directional couplers, quadrature hybrids, coupled-line directional coupler, and Lange coupler.

<u>RF Filters</u>: Included are the designs of lumped-element filters (Butterworth, Chebyshev, elliptic and Gaussian), filter transformations (low-pass, high-pass, band-pass, and band-stop), implementation of RF and microwave filters, low-pass filters (stepped-impedance and stub filters), band-pass filters (coupled-line and coupled-resonator filters), basic resonator and filter configurations, LC resonators, special filters, filter implementation, and coupled filters.

<u>Microwave Amplifiers</u>: Concepts include microwave transistors (bipolar & FET), gain and stability, noise linearity, input and output characteristics, noise floor, broadband amplifiers, multistage amplifiers, SFDR (spurious free dynamic range), BDR (blocking dynamic range), MDS (minimum detectable signal level), 1dB-compression point (input level at which the small-signal gain has dropped by 1 dB), IMD3 (third-order intermodulation), IIP3 (input-referred third-order intercept point), and OIP3 (output-referred third-order intercept point).

<u>Power Amplifiers</u>: Includes characteristics of power amplifiers, power amplifier classes, high-efficiency power amplifiers, large-signal impedance matching, and linearization techniques.

<u>Oscillators</u>: Included are basic LC oscillator topologies, voltage-controlled oscillators, phase noise in oscillators, and bipolar and CMOS LC oscillators.

<u>Microwave Circuit Fabrication Technologies</u>: Concepts include microwave printed circuits, hybrid microwave integrated circuits, monolithic integrated circuits, device technologies, CMOS fabrication, and fabrication by micromachining.

<u>Excitation and Coupling in RF and Microwave Circuits</u>: This includes aperture coupling, holes in waveguides, microstrip gaps and slots, ground-plane slots, proximity coupling, coupling via current loops, distributed coupling, hybrids, directional couplers, and coupled-line filters.

7.6   Instruments and Measurements [Wit02]

Wireless communication engineers must understand the key parameters for wireless system performance, as well as the operation of the instruments used to measure these parameters.

7.6.1 Basic Instruments and Measurements

Following are some basic measurement tools and concepts related to their proper use.

Electrical Measurement Principles: Concepts include measurement errors and accuracy, error estimation, noise types and effects, and instrument transformers and bridges.

Basic Electrical Instrumentation: Includes principles of operation, galvanometer, voltmeter, ammeter, wattmeter and power measurements, multimeters, and series and parallel connections.

Computer-controlled Measurements: Includes the IEEE-488 interface and associated software.

Signal Generators: Includes oscillators, RF and microwave sources, frequency, amplitude and power selection, internal/external AM/FM/PM/other advanced modulation, and tracking generators.

Oscilloscopes: Included are principles of operation, internal/external triggering, markers, frequency and amplitude measurement, X-Y display, and phase-angle measurements.

Time-domain Interferometers: Concepts included are TDR principles of operation, cable testing, applications to nondestructive testing, and fault testing of printed circuit boards with TDR.

Spectrum Analysis: Includes such concepts as basic functions (frequency and amplitude selection, bandwidth, span, markers, sweep and trace), measurement fundamentals (resolution and video bandwidth, sweep time, averaging), basic measurements (signal and channel power, occupied bandwidth, adjacent channel power, out-of-band spurious emissions, in-band/out-of-channel, carrier-to-interference ratio), interference measurements, measurement of specific signal types (AM, FM, SSB. GSM, CDMA, TDMA), and AM/FM demodulation.

Network Analysis: Includes such concepts as basic functions (frequency and amplitude selection, bandwidth, span, markers, sweep and trace), calibration (TRL and full two-port calibrations, connector types, calibration kits), transmission measurements (insertion loss and gain, 3-dB bandwidth, pass-band flatness, out-of-band rejection, phase response, electrical length, phase distortion, group delay), reflection measurements (return loss, reflection coefficient, SWR, impedance, admittance, Smith chart displays), and time domain measurements (time-domain reflection, gating time-domain response).

Protocol Analyzers: Concepts include principle of operation and BER measurements.

7.6.2 Power Calculations [Bea00]

Power calculations are a particularly important type of measurements for wireless transmission.

The Definition of Electric Power is the rate at which electric energy is dissipated by conversion into other forms of energy—electromagnetic, heat, sound and kinetic energy, etc. Power is most often measured in units of watts (W). One watt represents the electric power resulting from the dissipation of one joule (J) of electrical energy in one second.

Units of Power: These include Watts (W), ergs per second (erg/s), horsepower (hp), metric horsepower (PS), foot-pounds per minute (ft·lb/min), and Btu per hour (Btu/h); it is important to understand the conversion of and between units.

AC Power: Concepts include instantaneous, average, peak and real power (in W), reactive power (VAR), complex power, apparent power (in VA), and power factor.

Decibel Measurements: These include absolute (dBm, dBW, dBJ, dBf, dBk) or relative measurements (dBd, dBFS, dB-Hz, dBi, dBiC, dBO, dBrn, dBx, dBc and -dB, all values expressed relative to the carrier power).

7.7   Communication Networks

7.7.1 Digital Transmission and Switching [New06], [Wil06]

Wireless networks are parts of a larger worldwide telecommunications infrastructure, and it is important for the wireless engineer to be familiar with some of the aspects of this infrastructure.

Telephony: Concepts include the plain old telephone system (POTS), dual-tone, multi-frequency (DTMF) dialing, local loop, analog (voice-band) transmission, A-law/μ-law companding, pulse-code modulation, quantization and quantization noise, TDM, public-switched telephone network (PSTN), Signaling Systems (SS7), Integrated Services Digital Network (ISDN), and T and E carrier systems, the backbone of telephony networks (see Table 7-3). Other concepts include data communications over PSTN, X.25 and HDLC, voice-band modems, DSL (digital subscriber line) and its derivatives (ADSL, DSL++ and VDSL, etc.), cable modems, and DOCSIS (Data Over Cable Service Interface Specifications).

T-carrier and E-carrier Systems	North America	Europe (CEPT)*
Level zero (channel data rate)	64 kb/s (DS0)	64 kb/s
First level	1.544 Mb/s (DS1) (24 user channels) (T1)	2.048 Mb/s (32 user channels) (E1)
(Intermediate level, U.S. hierarchy only)	3.152 Mb/s (DS1C) (48 Ch.)	
Second level	6.312 Mb/s (DS2) (96 Ch.)	8.448 Mb/s (128 Ch.) (E2)
Third level	44.736 Mb/s (DS3) (672 Ch.) (T3)	34.368 Mb/s (512 Ch.) (E3)
Fourth level	274.176 Mb/s (DS4) (4032 Ch.)	139.264 Mb/s (2048 Ch.) (E4)
Fifth level	400.352 Mb/s (DS5) (5760 Ch.)	565.148 Mb/s (8192 Ch.) (E5)

*CEPT: European Conference of Postal and Telecommunications Administrations

Table 7-3: Specifications of T and E Carrier Systems

Additional Concepts: Includes baseband coding of waveforms, Manchester coding, return to zero (RZ) and non return to zero (NRZ) signaling, circuit switching, packet switching, slotted and pure Aloha, token ring, Ethernet, Carrier Sense Multiple Access with Collision Detection (CSMA/CD) and Collision

Avoidance (CSMA/CA), E1/T1 multiplexing, Optical Carrier OC-n, Synchronous Digital Hierarchy (SDH), Asynchronous Transfer Mode (ATM), fiber distributed data interface (FDDI), and frame relay.

Internet Protocol: Concepts include encapsulation, fragmentation and reassembly, connection control, ordered delivery, flow control and error control, IPV4 and IPV6, voice over IP, and IPTV. See section 7.7.3 for more detailed information about IP network concepts.

7.7.2 Basic Topics of Queuing Theory and Traffic Analysis [Hay04]

A wireless engineer should also have some fundamental knowledge related to traffic management on a communication network.

Combinatorial Analysis: Concepts include basic counting principles, permutations, combinations, sampling schemes, and binomial and multinomial coefficients.

Probability Models and Axioms: Includes sample spaces, events, and probability axioms.

Conditioning and Independence: Included are conditional probabilities, Bayes' Rule, and independent events.

Random Variables: Includes such concepts as discrete and continuous random variables, probability mass functions, expectations, and conditional expectation.

Probability Theory: In addition to random variables, this topic includes transformations, use of histograms, Markov chains and queuing theory; definition of traffic intensity and its measure, definition of queues, the Markovian case and the solution at regime, main queue analysis parameters, Little's theorem, M/M/s and similar queue analysis, the Erlang-B formula, and distribution of the queuing delay.

M/G/1 Queuing Theory and Applications: Concepts include solution of the state probability distribution for M/G/1 queues, different possibilities of embedding instants, differentiated service times, queuing theory applied to local area networks with analysis of token and polling schemes, analysis of different types of CSMA, and analysis of PRMA.

Additional Concepts: These include networks of queues and closed networks, traffic rate equations, the Burke theorem and Jackson theorem, with examples, Markovian models used for traffic generation: on/off sources, fluid flow traffic models, traffic loss analysis, multi-rate traffic analysis, traffic load measurement, sampling methods, traffic model selection, and traffic types and grades of service.

7.7.3 Internet Protocol (IP) Networks

This section describes some of the knowledge a wireless engineer should possess in connection with Internet Protocol transmission and IP networks in general.

The Internet: Concepts include Internet Protocol (IP), IPv4, IPv6, Transmission Control Protocol (TCP), User Datagram Protocol (UDP), IP address (netID and hostID), subnets, routing tables and adjunct protocols such as ARP (address resolution protocol) and ICMP (Internet control message protocol), and routing protocols.

IP Datagrams: IP is a connectionless protocol with all information contained in the payload of the datagram. The header of each datagram contains several fields, as shown in Table 7-4. Header length can vary from 5 (32-bit) words to 15 words as specified in the Intermediate Header Length (IHL).

0 Bits	4	8	16	19	31
Version (4)	Intermediate Header Length	Type of Service (TOS)	Total Length		
Identification			Flags	Fragment Offset	
Time to Live		Protocol	Header Checksum		
Source IP Address					
Destination IP Address					
Options + Padding					
Payload ≤ 65535 bytes					

Table 7-4: Datagram of Current IP Version 4 (IPv4)

IPv6: Devised to overcome limitations of IPv4, its main features include 128-bit address space, hierarchical addressing to size of the routing table, a simplified header, improved security, and harder quality-of-service guarantee. QoS guarantee is achieved through preferential treatment by routers of the packets associated with interactive and multimedia applications relative to those relating to traditional applications, such as email, and file transfers. Key differences can be seen in the datagrams for IPv4 (Table 7-4) and IPv6 (Table 7-5).

0 Bits	4	12	16	24	31
Version	Traffic Class	Flow Label			
Payload Length		Next Header		Hop Limit	
Source Address (16 octets)					
Destination Address (16 octets)					
Payload ≤ 65535 bytes					

Table 7-5: Datagram of Version 6 (IPv6)

7.7.3.1 Providing QoS in IP Networks

Applications and sessions utilize the IPv4 type of service (TOS) byte or IPv6 Class byte to signal their QoS needs to the networks. IPv6 Class bytes have the same meaning as TOS bytes. The Request for Change command (RFC 791) describes the usage of TOS bytes to guide the selection of the actual service parameters when a datagram is transmitted through a network. The 8 bits of the TOS field are illustrated in Figure 7-1.

```
        0     1     2     3     4     5     6     7
      +-----+-----+-----+-----+-----+-----+-----+-----+
      |     |     |     |     |     |     |     |     |
      |  PRECEDENCE     |  D  |  T  |  R  |  C  |  0  |
      |     |     |     |     |     |     |     |     |
      +-----+-----+-----+-----+-----+-----+-----+-----+


Bit   3:  0 = Normal delay        1 = Low delay
Bit   4:  0 = Normal throughput    1 = High throughput
Bit   5:  0 = Normal reliability   1 = High reliability
Bit   6:  0 = Normal cost          1 = Minimize cost


Precedence

    111 - Network control
    110 - Internetwork control
    101 - CRITIC/ECP
    100 - Flash override
    011 - Flash
    010 - Immediate
    001 - Priority
    000 - Routine
```

Figure 7-1: IP TOS Bytes as Defined in RFC 791

The network control precedence is intended to be used only within a network. RFC 791 recommends that the actual use and control of that designation be left to each network. The Internetwork control designation is intended for use only by gateway control originators.

DiffServ Codepoints and Explicit Congestion Notification ECN in IP TOS

RFC 2474 provides a replacement header field, called the DS field, which is intended to supersede the existing definitions of the IPv4 TOS octet and IPv6 Class byte. Six bits of the DS field are used as a codepoint (DSCP) to select the per-hop behaviors (PHB) that a packet experiences at each node. A two-bit ECT and CE field is used by the ECN framework, as shown in Figure 7-2. Values of the ECN bits are ignored by differentiated services-compliant nodes when determining the per-hop behavior to apply to a received packet. The DS field structure is presented in Figure 7-2.

```
        0   1   2   3   4   5   6   7
      +---+---+---+---+---+---+---+---+
      |         DSCP          | E | C |
      +---+---+---+---+---+---+---+---+

    DSCP: Differentiated services codepoint
     ECT: Explicit congestion notification (ECN)-capable transport
      CE: Congestion experienced
```

Figure 7-2: The DS Field Structure

In the DSCP value notation *xxxxxx* (where *x* may equal *0* or *1*) used here, the left-most bit signifies bit 0 of the DS field (in Figure 7-2), and the right-most bit signifies bit 5. The structure of the DS field is incompatible with the existing definition of the IPv4 TOS octet. The presumption is that DS domains protect themselves by deploying re-marking boundary nodes, as should networks using the RFC 791 precedence designations. This is done with the help of mapping tables or manual configuration as discussed below.

RFC 3168 recommends the use of ECN and CE bits to notify when there is congestion in the IP network. ECN and DiffServ (differentiated services) can work independently in a network. This RFC uses an ECN field in the IP header with two bits, making four ECN codepoints, *00* to *11*. ECN-capable transport (ECT) codepoints *10* and *01* are set by the data sender to indicate that the end-points of the transport protocol are ECN-capable. The not-ECT codepoint *00* indicates a packet that is not using ECN. The CE codepoint *11* is set by a router to indicate congestion at the end nodes. Routers that have a packet arriving at a full queue drop the packet, just as they do in the absence of ECN.

7.7.3.2 QoS Model

The QoS model specified here covers the features of DiffServ but not those of integrated services (IntServ). As presented in RFC 2475, DiffServ is a connectionless service that provides for service differentiation for aggregates of flows specified by the DiffServ codepoint (DSCP) field of the IP header.

DiffServ implementations provide both behavior aggregate (BA) (simple classification using only the DiffServ codepoint) and more complex multi-field (MF) classification, which is based on IP and layer 4 header fields. The classification functions of DiffServ are generally broken down into the following components:

- Policy and policy scope—provides a way to classify packets according to the policies of the operator.
- Hierarchical policy configuration specification—allows policies to be constructed of one or more classifications and actions. Classifications may be used by more than one policy.
- Policy actions—determine what is done once a packet is classified (drop, remark, police, etc.).
- Filtering—determines the inspection depth of the packet, that is, how far beyond the beginning of the packet it is inspected to determine what do with it.
- Mapping tables—allows the operator to map between packet designations, such as DSCP, precedence and CoS bits, to allow for remarking and administrative domain changes.

7.3.3.3 Queuing

Network nodes that implement the differentiated services enhancements to IP use a codepoint in the IP header to select a per-hop behavior (PHB). IETF has proposed two different PHB groups, expedited forwarding (EF) and assured forwarding (AF). The third PHB group (called best effort) has no service differentiation.

- Expedited forwarding (EF): Used to build a low-loss, low-latency, low-jitter, assured bandwidth, end-to-end service through DS domains.
- Assured forwarding (AF): The AF PHB group delivers IP packets in four independently forwarded AF classes (AF1, AF2, AF3, AF4). Within each AF class, an IP packet can be assigned one of three different levels of drop precedence.
- Best effort (BE): This PHB group acts like the normal IP with no consideration of the IP TOS byte.

Strict priority (SP) and weighted fair queuing (WFQ) scheduling mechanisms are used to schedule IP packets from these traffic class queues. The default router configuration does not give any forwarded traffic the network control class. If required, configuration of DiffServ policy or modification of DSCP trust and DSCP-to-class map table can be used to give forwarded network control traffic the same priority as the generated protocol traffic.

Active queue management (AQM) can be enabled for AF and BE traffic classes. Different AQM techniques monitor traffic load in an effort to anticipate and mitigate congestion at network bottlenecks. It is achieved through packet dropping. It will also be used to avoid global synchronization.

7.7.4 Frequency Allocations and Reuse

Frequency Bands [ITU-R vol.1]: Covers the various frequency bands and their use (low frequency, LF, 30 kHz to 300 kHz; very high frequency, VHF, 30 MHz to 300 MHz; ultrahigh frequency, UHF, 300 MHz to 3000 MHz; super high frequency, SHF, 3 GHz to 30 GHz; extremely high frequency, EHF, and 30 GHz to 300 GHz).

Frequency Bands Allocated to Wireless Communications Worldwide: Concepts include AMPS, GSM, 3G systems, IS-95, IS-136, SMR, wireless local area networks (WLANs), WiMAX, personal area networks (PANs), DECT, fixed/dynamic channel allocation, and radio resource management.

Cellular Concepts and Frequency Reuse: Iincludes cellular systems, frequency reuse factor and cell clusters, co-channel interference, adjacent channel interference, channel reuse ratio, power control, cell splitting, sectoring, cell coverage area, outage probability, cellular base station, hexagonal geometry, center-excited cells, edge-excited cells, frequency reuse factor in FDMA, CDMA and OFDMA networks, the 60-degree coordinate system, soft/hard handoff, handoff prioritization, and dwelling time.

Regulatory Bodies: Attention should be paid to the workings of the International Telecommunication Union (ITU) and the U.S. Federal Communications Commission (FCC).

7.8 Other Communication Systems

Wireless communication networks interface with numerous other communication systems, and a wireless engineer should be familiar with how these systems are designed and operate.

7.8.1 Basic Optical Communications

Fundamental concepts in optical communications include optical fibers, optical wave propagation, waveguide operation, single/multi-mode optical fiber, attenuation and dispersion in fiber, optical sources (lasers and LEDs), detectors (PIN and APD), WDM concepts, subcarrier multiplexing and analog transmission, and optical networks.

7.8.2 Basics of Satellites [ITU02, ETSI]

Satellite frequency bands, see Table 7-6.

Satellite orbits: Basic concepts include geosynchronous (GEO), medium-Earth orbit (MEO), and low-Earth orbit (LEO).

Band	Frequency	Service
L	1–2 GHz	Mobile
S	2.5–4 GHz	Mobile
C	3.7–8 GHz	Fixed
X	7.25–12 GHz	Military
Ku	12–18 GHz	Fixed
Ka	18–30.4 GHz	Fixed
V	37.5–50.2 GHz	Fixed

Table 7-6: Satellite Frequency Bands

Payload Architectures: Includes transparent bent pipe and onboard processing.

Performance Measures: Included are link availability, bit error rate, throughput, delay, and grade of service.

Rain Fade Mitigation Techniques: These include Uplink Power Control (UPC), Automatic Level Control (ALC), Adaptive Coding and Modulation (ACM), and site diversity. [ITU02]

Satellite Protocols: Included are Asynchronous Transfer Mode (ATM), Internet Protocol (IP), and ATM over IP.

User-Terminal Features: These include antennas, Low Noise Amplifiers (LNA), Power Amplifiers (PA), cost, and performance.

Interference Calculations: Concepts include co-channel interference from adjacent satellites, crosstalk interference, cross polarization interference, and cross modulation products.

Satellite Access: Includes Time, Frequency and Code Division Multiple Access (TDMA, FDMA, and CDMA) and contention schemes.

Transmitter Power: Includes types (Solid State Power Amplifier (SSPA) and Traveling Tube Power Amplifier (TWTA)) and performance (linearization and back-off).

Adaptive Coding Modulation Techniques: Concepts include channel coding and advanced modulation.

Antennas: Includes such concepts as types, gain, side-lobe performance, noise temperature, efficiency, tracking, and installation.

Losses: Types of losses include free space, pointing, absorption, rain fade, and depolarization.

Onboard Transponders and Power Amplifiers: Concepts include size (bandwidth and power) and performance.

Modem Performance: Includes such concepts as speed, error control, modulation efficiency, and performance threshold.

Satellite Services: Includes broadcast (radio, TV, data) and communications (VSAT, GPS, broadband).

Link Budget Calculations: Concepts include bandwidth and power utilization.

7.9   General Engineering Management and Economics

In addition to technical knowledge, a wireless communication engineer should possess management and decision-making skills and a fundamental understanding of engineering economics.

7.9.1 Fundamental Engineering Economics [New04]

The following topics in basic engineering economics should be familiar to a wireless engineer.

Basic Statistics Theory: Includes jointly distributed random variables, the weak law of large numbers, Central Limit theorem, the strong law of large numbers, mean, median, standard deviation, variance, range, inter-quartile range, confidence intervals, point estimation, test procedures, errors, and large sample test for population mean and population proportion, scatter plots, Pearson correlation coefficient, and simple linear regression. A number of these concepts should be familiar from their technical applications (identified earlier in this chapter); they also apply to economics.

Engineering Costs and Cost Estimation: Included are types of costs and cost estimates, models for cost estimation, and cash flow diagrams.

Interest and Equivalence: Concepts include types of interest, computing cash flows, equivalence calculations, and interest formulas.

Analysis Techniques: Many different types abound, including present worth, annual cash flow, rate of return, future worth, benefit-cost ratio, sensitivity, and break-even and replacement analyses.

Depreciation: This includes basic aspects of depreciation, depreciation methods, modified accelerated cost recovery system, asset disposal, and depletion.

Other important topics include taxation and after-tax economic analysis, inflation effects, capital budgeting, and accounting principles.

7.9.2 Project Management Methods and Processes [Gra02], [IEEE Std 1233-1998]

A wireless communication engineer should have an understanding of the following basic topics related to project management.

The Engineering Design Process: Includes prescriptive and descriptive processes, and elements of the design process.

Requirements Specification: Included are concept generation and evaluation, properties of engineering requirements, techniques for identifying requirements, properties of requirements specifications, requirements validation, and the engineering-marketing tradeoff matrix.

Functional Decomposition and Design: Key concepts include bottom-up and top-down design approaches, design architecture, functional specifications/requirements, state diagrams/state machines, flowcharts, data flow diagrams, entity relationship diagrams, and the Unified Modeling Language (UML).

Project Management: Includes work breakdown structures, network diagrams, the Gantt chart, cost estimation, elements of effective presentation, and reliability prediction.

Integration and Verification: Concepts include properties of test cases and units, and integration and acceptance tests.

7.9.3 *The IEEE Code of Ethics [IEEE CoE]*

As they go about their work, engineers should all be aware of the IEEE's Code of Ethics. It should guide them professionally in all they do. Membership in the IEEE in any grade carries the obligation to abide by the code. Here is the code in its entirety.

We, the members of the IEEE, in recognition of the importance of our technologies in affecting the quality of life throughout the world and in accepting a personal obligation to our profession, its members and the communities we serve, do hereby commit ourselves to the highest ethical and professional conduct and agree:

1. To accept responsibility in making decisions consistent with the safety, health and welfare of the public, and to disclose promptly factors that might endanger the public or the environment;

2. To avoid real or perceived conflicts of interest whenever possible, and to disclose them to affected parties when they do exist;

3. To be honest and realistic in stating claims or estimates based on available data;

4. To reject bribery in all its forms;

5. To improve the understanding of technology, it's appropriate application, and potential consequences;

6. To maintain and improve our technical competence and to undertake technological tasks for others only if qualified by training or experience, or after full disclosure of pertinent limitations;

7. To seek, accept, and offer honest criticism of technical work, to acknowledge and correct errors, and to credit properly the contributions of others;

8. To treat fairly all persons regardless of such factors as race, religion, gender, disability, age, or national origin;

9. To avoid injuring others, their property, reputation, or employment by false or malicious action;

10. To assist colleagues and co-workers in their professional development and to support them in following this code of ethics.

7.10   Key References

[Pro07] J.G. Proakis and M. Salehi, *Digital Communications, (5th ed.)*, McGraw-Hill Higher Education, 2007.

[Will06] W. Stallings, *Data and Computer Communications, (8th ed.)*, Prentice Hall, 2006.

[Whi03] J.F. White, *High Frequency Techniques: An Introduction to RF and Microwave Engineering*, John Wiley and Sons, 2003.

Appendices

Appendix A

Complete References & Further Resources

Chapter 1 – Wireless Access Technologies

"High speed downlink packet access (HSDPA); Overall description; Stage 2 (Release 5)," The 3rd Generation Partnership Project 2 (3GPP2), Tech. Rep. TS25.308 v5.7.0.,

"Physical channels and mapping of transport channels onto physical channels (FDD) (Release 5)," The 3rd Generation Partnership Project 2 (3GPP2), Tech. Rep., TS25.211 v5.8.0.

"Physical layer procedures (FDD)-Release 5," The 3rd Generation Partnership Project 2 (3GPP2), Tech. Rep. TS25.214 v5.11.0.

"FDD enhanced uplink; Overall description; Stage 2 (Release 6)," The 3rd Generation Partnership Project 2 (3GPP2), Tech. Rep. TS25.309 v6.6.0.

"Feasibility Study for enhanced uplink for UTRA FDD," The 3rd Generation Partnership Project 2 (3GPP2), Tech. Rep.TR25.896 v6.0.0.

Jin Yang et al, "Design aspects and system evaluations of IS-95 based CDMA system," *Proc. Internal Conf. Universal Personal Communications,* pp.381–385, Oct. 1997.

Jin Yang and Jinsong Lin, "Optimization of power management in a CDMA radio network," *Proc. IEEE Vehic. Technol. Conf.,* Sept. 2000.

"Mobile station—Base station compatibility standard for wideband spread spectrum cellular system," ANSI/TIA/EIA-95-B, Tech Rep.; Feb. 1999.

"cdma2000 Standards for Spread Spectrum Systems," TIA/EIA/IS-2000-D, Feb. 2004.

"cdma2000 High Rate Packet Data Air Interface Specification," The 3rd Generation Partnership Project 2 (3GPP2), Tech. Rep. C.S0024-A, July 2005.

Jin Yang, "Performance and Deployment of a Mobile Broadband Wireless Network Based on IS-856 (EV-DO)," *Proc. 2004 World Wireless Congress,* pp. 398–402.

Jin Yang, "Multimedia Services Over 1xEV-DO Mobile Broadband Network," *Proc. 2006 SATEC.*

Rainer Bachl et al., "The Long Term Evolution Towards a New 3GPP Air Interface Standard," *Bell Labs Tech. J.,* vol. 11, Issue 4, pp. 25, Winter 2007.

"Requirements for evolved UTRA and UTRAN," The 3rd Generation Partnership Project, Tech. Rep. TR25.913 V7.3, March 2006.

Suman Das et al., "EV-DO revision C: Evolution of the cdma2000 Data Optimized System to Higher Spectral Efficiencies and Enhanced Services, *Bell Labs Tech. J.*, vol. 11, issue 4, pp. 5–24, Winter 2007.

"Mobile WiMAX–Part I: A technical overview and performance evaluation," WiMAX Forum, Feb. 2006.

"Mobile WiMAX–Part II: Competitive analysis," WiMAX Forum, Feb. 2006.

Hassan Yagoobi, "Scalable OFDMA physical layer in IEEE 802.16 Wireless MAN," Intel Technol. J., vol. 08, August 2004.

G. Nair et al., "IEEE 802.16 medium access control and service provisioning," Intel Technol. J., vol. 08, Aug. 2004.

International Telecommunication Union, Radiocommunication Sector, "Framework and overall objectives of the future development of IMT-2000 and systems beyond IMT-2000," ITU-R Rec. M.1645, June 2003.

R. Tafazolli (Ed.), *Technologies for the Wireless Future, Wireless World Research Forum (WWRF)*, vol. 2, John Wiley & Sons, 2006.

http://www.wireless-world-research.org/

http://www.ngmn.org

https://www.ist-winner.org

Next Generation Mobile Communications Forum (NGMC Forum), http://www.ngmcforum.org

Defense Advanced Research Projects Agency (DARPA). "NeXt generation (XG) communications program." [Online]. Available: AV http://www.darpa.mil/sto/smallunitops/xg.html

A. Alexiou and D. Falconer, "Challenges and trends in the design of a new air interface," *IEEE Vehic. Technol. Mag.*, vol. 1, no. 2, pp. 16–23, 2006

A. Alexiou et al, "Duplexing, resource allocation and inter-cell coordination-design recommendations for next generation systems," *Wireless Commun., and Mobile Computing Mag.*, vol. 5, p.77–93, 2005.

R. Tafazolli, Ed., *Technologies for the Wireless Future, Wireless World Research Forum (WWRF)*, vol. 1, John Wiley & Sons, 2005.

C. Antón et al, "Cross-Layer scheduling for multi-user MIMO systems, *IEEE Commun, Mag.*, vol. 44, No. 9, pp. 39–45, Sept. 2006.

Chapter 2 - Network and Service Architecture

[Var2003] V.K. Varma, K.D. Wong, K.-C. Chaing, and F. Paint, *IEEE Commun, Mag.* special issue on "Integration of wireless LAN and 3G wireless technologies," November 2003.

[Won2003] K.D. Wong and V.K. Varma, "Supporting real time IP multimedia services in UMTS," *IEEE Commun. Mag.*, Nov. 2003.

[Won2005] K.D. Wong, "The IP multimedia sub-system," chapter 12 of *Wireless Internet Telecommunications,* Artech House, 2005.

[Come2006] D. Comer, *Internetworking With TCP/IP Volume 1: Principles Protocols, and Architecture*, 5th ed., Prentice Hall, 2006.

[Deer1998] S. Deering and R. Hinden, "Internet Protocol, version 6 (IPv6) specification," RFC 2460, Dec. 1998.

[Malk1998] G. Malkin, "RIP Version 2," RFC 2453, Nov. 1998.

[Moy1998] J. Moy, "OSPF Version 2," RFC 2328, April 1998.

[Colt1999] R. Coltun, D. Ferguson, and J. Moy, "OSPF for IPv6," RFC 2740, Dec. 1999.

[ISO2002] ISO/IEC 10589, "Information Technology; Telecommunications and Information Exchange Between Systems; Intermediate System to Intermediate System Intra-Domain Routing Information Exchange Protocol for Use in Conjunction with the Protocol for Providing the Connectionless-Mode Network Service (ISO 8473)," 2nd ed., 2002.

[Perk1996] C. Perkins (Ed.), "IP Mobility Support," RFC 2002, Oct. 1996.

[Deep2005] V. Devarapalli, R. Wakikawa, A. Petrescu, and P. Thubert, "Network Mobility (NEMO) Basic Support Protocol," RFC 3963, Jan. 2005.

[John2004] D. Johnson, C. Perkins, and J. Arkko, "Mobility Support in IPv6," RFC 3775, June 2004.

[3GPP29.060] 3GPP TS 29.060, "General Packet Radio Service (GPRS); GPRS Tunnelling Protocol (GTP) Across the Gn and Gp Interface."

[Ros2002] J. Rosenberg, H. Schulzrinne, G. Carmarillo, A. Johnston, J. Peterson, R. Sparks, M. Handley, and E. Schooler, "SIP: Session Initiation Protocol," RFC 3261, July 2002. http://www.rfceditor.org/rfc/rfc3261.txt

[Schulzrinne-SIP] H. Schulzrinne's SIP site, http://www.cs.columbia.edu/~hgs/sip

[Sch2003] H. Schulzrinne, S. Casner, R. Frederick, and V. Jacobson, "RTP: A Transport Protocol for Real-Time Applications," RFC 3550, July 2003. ftp://ftp.rfc-editor.org/in-notes/rfc3550.txt

[Schulzrinne-RTP] H. Schulzrinne's RTP site, http://www.cs.columbia.edu/~hgs/rtp

[Bor2001] C. Bormann, C. Burmeister, M. Degermark, H. Fukushima, H. Hannu, L-E. Jonsson, R. Hakenberg, T. Koren, K. Le, Z. Liu, A. Martensson, A. Miyazaki, K. Svanbro, T. Wiebke, T. Yoshimura and H. Zheng, "Robust Header Compression (ROHC): Framework and Four Profiles: RTP, UDP, ESP, and Uncompressed," July 2001. http://tools.ietf.org/html/rfc3095

[Cor2006] Carlos Cordeiro and Dharma P. Agrawal, *Ad hoc & Sensor Networks: Theory and Applications*, World Scientific Publishing, Spring 2006. http://www.worldscibooks.com/engineering/6044.html

[Gup2000] P. Gupta and P.R. Kumar, "The capacity of wireless networks," *IEEE Trans. on Information Theory*, vol. 46(2), 2000.

[Bej2004] Y. Bejerano, "Efficient integration of multihop wireless and wired networks with QoS constraints," *IEEE/ACM Trans. on Networking*, vol 12(6), pp. 1064–1078, 2004.

[Pra2006] R. Prasad and H. Wu, "Minimum-Cost Gateway Deployment in Cellular WiFi Networks," *Proc. Consumer Communications & Networking Conf.*, Las Vegas, Jan. 2006.

[He2007] B. He, B. Xie, and D. P. Agrawal, "Optimizing the Internet gateway deployment in a wireless mesh network," *Proc. Fourth IEEE International Conf. on Mobile Ad-hoc and Sensor Systems (MASS)*, Pisa, Italy, Oct. 2007.

[Wan2007] J. Wang, B. Xie, K. Cai, and D.P. Agrawal, "Efficient mesh router placement in wireless Mesh networks," *Proc. Fourth IEEE International Conf. on Mobile Ad-hoc and Sensor Systems (MASS)*, Pisa, Italy, Oct. 2007.

[Nan2007] D. Nandiraju, N. Nandiraju, and D.P. Agrawal, "Service differentiation in IEEE 802.11 mesh networks: a dual queue strategy," MILCOM, Orlando, FL., Oct. 29–31, 2007.

[Nan2007a] N.S. Nandiraju, D. Nandiraju, L. Santhanam and D.P. Agrawal, "A cache based traffic regulator for improving performance in IEEE 802.11s-based mesh networks," *IEEE Radio and Wireless Symp. (RWS 2007)*, Long Beach, CA., Jan. 2007.

[San2006] L. Santhanam, N. Nandiraju, Y. Yoo, and D.P. Agrawal, "Distributed self-policing architecture for fostering node cooperation in wireless mesh networks," *Personal Wireless Commun.*, Sept. 20–22, 2006.

[Per2003] C. Perkins, E. Belding-Royer, and S. Das, "RFC3561: Ad hoc On-demand Distance Vector (AODV) Routing," The Internet Engineering Task Force (IETF), 2003.

[Joh2001] D.B. Johnson, D.A. Maltz, and J. Broch, "DSR: The Dynamic Source Routing Protocol for Multi-Hop Wireless Ad Hoc Networks," *Ad Hoc Networking*, vol. 1, C.E. Perkins, Ed.: Addison-Wesley, pp. 139–172, 2001.

[Per1994] C.E. Perkins and P. Bhagwat, "Highly dynamic destination-sequenced distance-vector routing (DSDV) for mobile computers," *Proc. ACM SIGCOMM*, pp. 234–244, 1994.

[Cla2003] T. Clausen and P. Jacquet, "RFC3626: Optimized Link State Routing Protocol (OLSR)," The Internet Engineering Task Force (IETF), 2003.

[Pea1999] M. R. Pearlman and Z. J. Haas, "Determining the optimal configuration for the zone routing protocol," *IEEE J, of Selected Areas in Commun.*, vol. 17, pp. 1395–1414, 1999.

[Ko2000] Y.B. Ko and N.H. Vaidya, "Location-aided routing (LAR) in mobile ad hoc networks," *Wireless Networks*, vol. 6, pp. 307–321, 2000.

[Kle1975] L. Kleinrock, *Queuing Systems; Vol. 1: Theory*, John Wiley & Sons, 1975.

[Chle1995] E. Chlebus and W. Ludwin, "Is handoff traffic really Poissonian?" *IEEE International Conf. on Universal Personal Commun. (ICUPC)*, pp. 348–353, Nov. 1995.

[Fang1999] Y. Fang and I. Chlamtac, "Teletraffic analysis and mobility modeling for PCS networks," *IEEE Trans. on Commun.*, vol. 47, no. 7, pp. 1062–1072, July 1999.

[Sidi1996] M. Sidi and D. Starobinski, "New call blocking versus handoff in cellular networks," *INFOCOM*, 1996.

[Oh1992] S. Oh and D. Tcha, "Prioritized channel assignment in a cellular radio network," *IEEE Trans. on Commun.*, vol. 40, no. 7, pp. 1259–1269, July 1992.

[Oliv1999] M. Oliver and J. Borras, "Performance evaluation of variable reservation policies for hand-off prioritization in mobile networks," *INFOCOM*, 1999, pp. 1187–1194.

[Choi2000] S. Choi and K. Sohraby, "Analysis of a mobile cellular system with hand-off priority and hysteresis control," *INFOCOM*, pp. 217–224, 2000.

[Katz1996] I. Katzela and M. Naghshineh, "Channel assignment schemes for cellular mobile telecommunication systems: a comprehensive survey," *IEEE Personal Commun.*, vol. 3(3), pp. 10–31, June 1996.

[Ekic2001] E. Ekici and C. Ersoy, "Multi-tier cellular network dimensioning," *Wireless Networks*, vol. 7(4), pp. 401–441, 2001.

[Orti1997] L. Ortigoza-Guerrero and A. Aghvami, "On the optimum spectrum partitioning in a microcell/macrocell cellular layout with overflow," *Globecom*, Phoenix, AZ, pp. 991–995, 1997.

[Kauf1981] L. Kaufman, B. Gopinath, and E. Wunderlich, "Analysis of packet network congestion control using sparse matrix algorithms," *IEEE Trans. on Commun.*, vol. 29, no. 4, pp. 453–465, Apr. 1981.

[Zhou1998] Y. Zhou and B. Jabbari, "Performance modeling and analysis of hierarchical wireless communications networks with overflow and take-back traffic," in PIMRC, pp. 1176–1180, 1998.

C. Perkins, *Mobile IP: Design Principles and Practices*, Prentice Hall, 1998.

H. Soliman, *Mobile IPv6: Mobility in a Wireless Internet*, Addison-Wesley Professional, 2004.

J. Solomon, *Mobile IP: the Internet Unplugged*, Prentice Hall, 1997.

M. Blanchet, *Migrating to IPv6: A Practical Guide to Implementing IPv6 in Mobile and Fixed Networks*, John Wiley & Sons, 2006.

K.D. Wong, *Wireless Internet Telecommunications*, Artech House, 2005.

D. Collins, *Carrier Grade Voice Over IP*, 2nd ed., McGraw-Hill, 2002.

J. Davidson, J. Peters, M. Bhatia, S. Kalidindi, and S. Mukherjee, *Voice over IP Fundamentals*, 2nd ed., Cisco Press, 2006.

M.A. Miller, *Voice over IP Technologies: Building the Converged Network*, Hungry Minds, 2002.

D. Minoli and E. Minoli, *Delivering Voice over IP Networks*, 2nd ed., John Wiley & Sons, 2002.

J. Rosenberg and H. Schulzrinne, "Reliability of Provisional Responses in Session Initiation Protocol (SIP)," June 2002, ftp://ftp.rfc-editor.org/in-notes/rfc3262.txt

J. Rosenberg and H. Schulzrinne, "Session Initiation Protocol (SIP): Locating SIP Servers," June 2002, ftp://ftp.rfc-editor.org/in-notes/rfc3263.txt

J. Rosenberg and H. Schulzrinne, "An Offer/Answer Model with Session Description Protocol (SDP)," June 2002, ftp://ftp.rfc-editor.org/in-notes/rfc3264.txt

A.B. Roach, "Session Initiation Protocol (SIP)-Specific Event Notification," June 2002, ftp://ftp.rfc-editor.org/in-notes/rfc3265.txt

S. Olson, G. Camarillo, and A. B. Roach, "Support for IPv6 in Session Description Protocol (SDP)," June 2002, ftp://ftp.rfc-editor.org/in-notes/rfc3266.txt

J. Varga et al, "Signaling Flows for the IP Multimedia Call Control Based on Session Initiation Protocol (SIP) and Session Description Protocol (SDP); Stage 3," 2005–2006. http://www.3gpp.org/ftp/Specs/html-info/24228.htm

K. Drage et al, "Internet Protocol (IP) Multimedia Call Control Protocol Based on Session Initiation Protocol (SIP) and Session Description Protocol (SDP); Stage 3 (Release 7)," 2005–2006. http://www.3gpp.org/ftp/Specs/html-info/24229.htm

K. Drage et al, "Presence Service Based on Session Initiation Protocol (SIP); Functional Flows, Information Flows, and Protocol Details (Release 6)," 2004, http://www.3gpp.org/ftp/Specs/html-info/24841.htm

F. Akyildiz, X. Wang, and W. Wan, "Wireless mesh networks: a survey," *Computer Networks and ISDN Systems*, vol. 47, no. 4, 2005.

Nagesh Nandiraju, Deepti Nandiraju, Lakshmi Santhanam, Bing He, Junfang Wang, and Dharma P. Agrawal, "Wireless mesh network: current challenges and future directions of web-in-the-sky," *IEEE Wireless Commun.*, vol. 14, no. 4, pp. 2–12, Aug. 2007,

A. Raniwala, K. Gopalan, and T. Chiueh, "Centralized channel assignment and routing algorithms for multi-channel wireless mesh networks," *Mobile Computing and Commun, Rev.*, vol. 8, no. 2, 2004.

Bin Xie, Yingbing Yu, Anup Kumar, and Dharma P. Agrawal, "Load-balanced mesh router migration for wireless mesh networks," *J. Parallel and Distributed Computing*, 2008.

E. Belding-Royer and C. K. Toh, "A review of current routing protocols for ad hoc mobile wireless networks," *IEEE Personal Commun.*, vol. 6, pp. 46–55, 1999.

J. Broch, D. A. Maltz D. B. Johnson, Y. C. Hu, and J. Jetcheva, "A performance comparison of multi-hop wireless ad hoc network routing protocols," *Proc. 1998 ACM/IEEE International Conf. Mobile Computing and Networking*, pp. 85–97.

IETF MANET Working Group, http://www.ietf.org/html.charters/manet-charter.html

Michel Mouly and Marie-Bernadette, *The GSM System for Mobile Communications*, Telecom Publishing, June 1992.

Theodore S. Rappaport, *Wireless Communications: Principles and Practice*, 2nd ed., Prentice Hall PTR, Jan. 2002.

Abbas Jamalipour, *The Wireless Mobile Internet: Architectures, Protocols and Services*, John Wiley & Sons, 2003.

Yi-Bing Lin and Imrich Chlamtac, *Wireless and Mobile Network Architectures*, (1st ed., paperback), John Wiley & Sons, 2000.

Gonzola Camarillo and Miguel-Angel Garcia-Martin, *The 3G IP Multimedia Subsystem (IMS): Merging the Internet and the Cellular Worlds*, 2nd ed., John Wiley & Sons, 2006.

Miikka Poiselka, Aki Niemi, Hisham Khartabil, and Georg Mayer, *The IMS: IP Multimedia Concepts and Services*, 2nd Edition, John Wiley & Sons, Mar. 2006.

Colin Perkins, *RTP: Audio and Video for the Internet*, Addison-Wesley Professional, June 2003.

Henry Sinnreich and Alan B. Johnston, *Internet Communications Using SIP: Delivering VoIP and Multimedia Services with Session Initiation Protocol (Networking Council)*, 2nd ed., John Wiley & Sons, 2006.

Chapter 3 – Network Management and Security

[Arba2001] W. Arbaugh et al., "Your 802.11 wireless network has no clothes," *IEEE Wireless Commun. Mag.*, vol. 6, pp. 44–51, Dec. 2002. <<JEAN: I GOOGLED THE TITLE

[Biha2005] E. Biham, O. Dunkelman, N. Keller, *A Related-Key Rectangle Attack on the Full KASUMI*, ASIACRYPT 2005, pp 443–461.

[Colb1997] B. Colbert, "On the Security of Cryptographic Algorithms," Ph.D. dissertation, University of University of New South Wales, 1997.

[Cour2002] N. Courtois, J. Pieprzyk, "Cryptanalysis of block ciphers with overdefined systems of equations," ASIACRYPT 2002, *8th International Conf. on Theory and Application of Cryptology and Information Security*, Queenstown, New Zealand, pp. 267–287.,2002.

[ISO 27002] ISO/IEC 27002, "Information Technology–Security Techniques–Code of Practice for Information Security Management."

[IEEE 802.11i 2004] "IEEE Standard for Information Technology–Telecommunications and Information Exchange Between Systems–Local and Metropolitan Area—Specific Requirements." "Part 11 Wireless LAN Medium Access Control (MAC) and Physical Layer (PHY) Specifications Amendment 6: Medium Access Control (MAC) Security Enhancements," IEEE Computer Society.

[Køie2002] G.M. Køien, "A Validation Model of the UMTS Authentication and Key Agreement Protocol," Technical Report R&D N 59/2002, Telnor, Dec. 2002.

[Lehe2006] G. Lehembre, "Seguridad WiFi–WEP, WPA, WPA2", *hakin9 Mag.*, no. 1, 2006.

[NIST SP800-48] National Institute of Standards and Technology, "Wireless Network Security for IEEE 802.11a/b/g and Bluetooth."

[NIST SP800-97] National Institute of Standards and Technology, Special Publication, *Establishing Wireless Robust Security Networks: A Guide to IEEE 802.11i*.

[OGC 2007] *The Official Introduction to the ITIL Service Lifecycle*, Office of Government Commerce, 2007.

[OGC 2007 2], *Service Transition*, Office of Government Commerce, 2007.

[RFC 1155] *Structure and Iidentification of Managemen Information for TCP/IP-based Iinternets*, The Internet Engineering Task Force.

[RFC 1157] *Simple Network Management Protocol (SNMP)*, The Internet Engineering Task Force.

[RFC 2865] *Remote Authentication Dial In User Service (RADIUS)*, The Internet Engineering Task Force.

[RFC 3584] *Coexistence between Version 1, Version 2, and Version 3 of the Internet-standard Network Management Framework*, The Internet Engineering Task Force.

[RFC 3588] *Diameter Base Protocol,* The Internet Engineering Task Force.

[RFC 3748] *Extensible Authentication Protocol (EAP),* The Internet Engineering Task Force..

[TMF 2007] "Enhanced Telecom Operations Model (eTOM)–The Business Process Framework, "TeleManagement Forum, 2007.

[TS 33.120] 3GPP Technical Specification Group Services and System Aspects; Security Principles and Objectives.

[TS 33.105] 3GPP Technical Specification Group Services and System Aspects; Cryptographic Algorithm Requirements.

[Zel99] D. Zeltserman, *A Practical Guide to SNMPv3 and Network Management*, Prentice Hall, 1999.

Chapter 4 – Radio Frequency Engineering, Propagation and Antennas

Antennas

[And07] Andrew Corp. "Basic antenna types used in cellular style systems," *Base Station Antenna Systems, Product Selection Guide 2007* http://www.andrew.com/search/docviewer.aspx?docid=7871

[Cal06] G. Calcev and M. Dillon, "Antenna tilt control in CDMA networks," *2nd Annual International Wireless Internet Conf.*, Boston, Aug. 2006.

[Con04] A. Constantinides and A. Schacham. (2004) *MIMO Wireless Systems* [Online]. Available at http://www.columbia.edu/~acc40/MIMO.pdf

[Die00] C.B. Dietrich, Jr., "Adaptive arrays and diversity antenna configurations for handheld wireless communication terminals," Ph.D. dissertation, Virginia Tech, Blacksburg, VA, 2000.

[FCC08] http://www.fcc.gov/oet/info/rules/part15/part15-9-20-07.pdf

[Gie02] J. Geier (2002), *RF Site Survey Steps* [Online]. Available at http://www.wi-fiplanet.com/tutorials/article.php/1116311

[Ges03] Dept. of Informatics, University of Oslo (2003, Sept. 26), "MIMO space-time coded wireless systems," [Online]. Available at http://www.iet.ntnu.no/projects/beats/Documents/GesbertMIMOlecture.pdf

[Han98] R.C. Hansen, *Phased Array Antennas*, John Wiley & Sons, 1998.

[Hea08] R. Heath, *Antenna Design And Analysis For MIMO Communication Systems.* [Online]. Available at http://users.ece.utexas.edu/~rheath/research/mimo/antennas

[Hou04] H. Hourani, "An Overview of Diversity Techniques in Wireless Communication Systems," Helsinki University of Technology, S-72.333 Postgraduate Course in Radio Communications, 2004/2005.

[IEEE83] *IEEE Standard Definitions of Terms for Antennas*, IEEE Std 145-1983.

[Kra02] J.D. Kraus, *Antennas,* 3rd ed., McGraw-Hill, 2002.

[Pet04] M. Pettersen et al, "Automatic antenna tilt control for capacity enhancement in UMTS FDD," *IEEE Vehicular Technol. Conf.*, vol. 1, pp. 26–29, 280–284, Sept. 2004.

[Rap02] T. Rappaport, *Wireless Communications,* 2nd ed., Prentice Hall, 2002.

[Sio05] I. Siomina, "P-CPICH power and antenna tilt optimization in UMTS networks," *Proc. IEEE Advanced Industrial Conf. on Telecommun.,* 2005.

[Vol07] J.L. Volakis, *Antenna Engineering Handbook,* 4th ed., McGraw-Hill, 2007.

[Wer00] D. Werner and R. Mittra, *Frontiers in Electromagnetics.* IEEE Press, 2000.

Propagation

Ber87] R.C. Bernhardt, "Macroscopic diversity in frequency reuse systems," *IEEE J. Select. Areas Commun.,* vol. SAC 5, pp. 862–878, June 1987.

[Bue04] R.M. Buehrer et al, "Ultra-wideband propagation measurements and modeling," DARPA NETEX Program and Virginia Tech, Final Rep., 31 Jan. 2004. Available at http://www.mprg.org/people/buehrer/ultra/darpa_netex.shtml

[Bul47] K. Bullington, "Radio propagation at frequencies above 30 megacycles," *Proc. IEEE,* vol. 35, pp. 1122–1136, 1947.

[Bul77] K. Bullington, "Radio propagation for vehicular communications," *IEEE Trans. Vehic. Technol.,* vol. VT 26, pp. 295–308, Nov. 1977.

[Cox83] D.C. Cox et al, "Measurements of 800-MHz radio transmission into buildings with metallic walls," *Bell Sys. Technic. J..,* vol. 62, no. 9, pp. 2695–2717, Nov. 1983.

[Cox84] D.C. Cox et al., "800 MHz attenuation measured in and around suburban houses," *AT&T Bell Lab. Technic. J.,* vol. 673, no. 6, July–Aug. 1984.

[Dev90] D.M.J. Devasirvatham et al., "Multi-frequency radiowave propagation measurements in the portable radio environment," *1990 IEEE International Conf. Commun.,* pp. 1334–1340.

[Dey66] J. Deygout, "Multiple knife-edge diffraction of microwaves," *IEEE Trans. Antennas and Propag.,* vol. AP-14, no. 4, pp. 480–489, 1966.

[Egl57] J.J. Egli, "Radio propagation above 40 Mc over irregular terrain, " *Proc. IEEE,* vol. 45(10), pp. 1383–9.

[Eps53] J. Epstein and D.W. Peterson, "An experimental study of wave propagation at 840 M/C," *Proc. IRE,* vol. 41, no. 5, pp. 595–611, 1953.

[Gri87] J. Griffiths, *Radio Wave Propagation and Antennas,* Prentice Hall International, 1987.

[Hat90] M. Hata, "Empirical formula for propagation loss in land mobile radio services," *IEEE Trans. Vehic. Technol.,* vol. VT-29, no. 3, pp. 317–325, Aug. 1990.

[Jak74] W.C. Jakes, Jr., *Microwave Mobile Communication,* Wiley-Interscience, 1974.

[Kra50] J.D. Krauss, *Antennas,* McGraw-Hill, 1950.

[Kre94] P. Kreuzgruber et al., "Prediction of indoor radio propagation with the ray splitting model including edge diffraction and rough surfaces," *Proc. 1994 IEEE Vehic. Technol. Conf.,* Stockholm, pp. 878–882.

[Lee85] W.C.Y. Lee, *Mobile Communications Engineering,* McGraw-Hill, 1985.

[McK91] J.W. McKeown and R.L. Hamilton, "Ray-tracing as a design tool for radio networks," *IEEE Networks Mag.*, vol. 5(6), pp. 27–50, 1991.

[Meh94] A. Mehrotra, *Cellular Radio Performance Engineering*, Artech House, 1994.

[Mil62] G. Millington et al., "Double knife-edge diffraction in field strength predictions," *Proc. IEE*, no. 109C, pp. 419–29, 1962.

[Mor00] R.K. Morrow and T.S. Rappaport, (2000, Mar. 1) "Getting In," *Wireless Rev. Mag.*, pp. 42–44. Available at *http://wirelessreview.com/issues/2000/0030l/feat24.htm*

[Oku68] T. Okumura et al., "Field strength and its variability in VHP and UHF land mobile service," *Rev. Electrical Commun. Labs*, vol. 16. no. 9, 10, pp. 825–873, Sept.–Oct. 1968.

[Rap91] T.S. Rappaport, "The Wireless Revolution," *IEEE Commun. Mag.*, pp. 52–71, Nov. 1991.

[Rap02] T.S. Rappaport, *Wireless Communications*, 2nd ed, Prentice Hall, 2002.

[Reu74] D.O. Reudink, "Properties of Mobile Radio Propagation Above 400 MHz," *IEEE Trans. Vehic. Technol.*, vol. 23, no. 2, pp. 1–20, Nov. 1974.

[Rus93] T.A. Russel et al, "A deterministic approach to predicting microwave diffraction by buildings for microcellular systems," *IEEE Trans. Antennas Propag.*, vol. 41, no. 12, pp. 1640–1649, Dec. 1993.

[Sch92] K.R. Schaubach et al., "A ray tracing method for predicting path loss and delay spread in microcellular environments," *42nd IEEE Vehic. Technol. Conf.*, Denver, 1992, pp. 932–935.

[Sch94] K.R. Schaubach and N.J. Davis, "Microcellular radio-channel propagation prediction, *IEEE Antennas Propag. Mag.*, 1994, 36(4), pp. 25–34.

[Sei91] S.Y. Seidel et al., "Path loss, scattering and multipath delay statistics in four European cities for digital cellular and microcellular radiotelephone," *IEEE Trans. on Vehic. Technol.*, 1991, vol. 40. no. 4, pp. 721–730.

[Sei92] S.Y. Seidel and T.S. Rappaport, "914-MHz path loss prediction models for indoor wireless communications in multi-floored buildings," *IEEE Trans. Antennas Propag.*, 1992, vol. 40. no. 2. pp. 2H–2I7.

[Sei94] S.Y. Seidel and T.S. Rappaport, "Site-specific propagation prediction for wireless in-building personal communication system design," *IEEE Trans Vehic. Technol.*, 1994, vol. 43, no. 4.

[Tur87] A.M.D. Turkmani et al., "Radio propagation into buildings at 441, 900, and 1400 MHz," *Proc. 4th International Conf. on Land Mobile Radio*, 1987.

[Tur92] A.M.D. Turkmani and A.F. Toledo, "Propagation into and within buildings at 900, 1800, and 2300 MHz," *1992 IEEE Vehic. Technol. Conf.*

[Val93] R.A. Valenzuela, "A ray tracing approach to predicting indoor wireless transmission," *Proc. IEEE Vehic. Technol. Conference*, 1993, pp. 214–218.

[Vio88] E.J. Violette et al. "Millimeter-wave propagation characteristics and channel performance for urban-suburban environments," National Telecommunications and Information Administration, NTIA Report 88-239, Dec. 1988.

[Wag94] J. Wagen and K. Rizk, "Ray tracing based prediction of impulse responses in urban microcells," *Proc. 1994 IEEE Vehic. Technol. Conf.*, Stockholm, pp. 210–214.

[Whi88] J.H. Whitteker, "Measurements of path loss at 910 MHz for proposed microcell urban mobile systems," *Proc. 1988 IEEE Trans. Vehic. Technol.*, vol. 37, no. 3, pp. 125–129.

RF Engineering

[Bli76] H.J. Blinchikoff and A.I. Zverev, *Filtering in the Time and Frequency Domains*, Robert E. Krieger Publishing, 1976.

[Cou93] L.W. Couch, *Digital and Analog Communications Systems*, Macmillan, 1993.

[Gos86] W. Gosling, *Radio Receivers*, Peter Peregrinus Ltd., 1986.

[Hay03] W. Hayward et al., *Experimental Methods in RF Design*, ARRL, 2003.

[Hay04] W. Hayward, *Introduction to Radio Frequency Design*, ARRL, 2004.

[Maa92] S. Maas, *Microwave Mixers*, 2nd ed., Artech House, 1992.

[Roh88] U.L. Rohde and T.T.N. Bucher, *Communications Receivers, Principles and Design*, McGraw-Hill, 1988.

[Sab87] W.E. Sabin and E.O. Schoenike, *Single-Sideband Systems and Circuits*, McGraw-Hill, 1987.

[Zve67] A.I. Zverev, *Handbook of Filter Synthesis*, John Wiley & Sons, 1967.

C. Balanis, *Advanced Engineering Electromagnetics*. John Wiley & Sons, 1989.

C. Balanis, *Antenna Theory, Analysis and Design*, 3rd ed. Hoboken, NJ: John Wiley & Sons, 2005.

E. Biglieri et al., *MIMO Wireless Communications*, Cambridge University Press, 2007.

H. Bölcskei et al., *Space-Time Wireless Systems: From Array Processing to MIMO Communications*, Cambridge University Press, 2006.

D. Dobkin, *RF Engineering for Wireless Networks: Hardware, Antennas, and Propagation*, Newnes Publishing, 2004.

R. Elliott, *Antenna Theory and Design*, IEEE Press, 2003.

F. Gross, *Smart Antennas for Wireless Communications*, McGraw-Hill, 2005.

R.C. Hansen, *Phased Array Antennas*, John Wiley & Sons, 1997.

R. Harrington, *Time Harmonic Electromagnetic Fields*, McGraw-Hill, 1961.

J.D. Kraus, *Antennas*, 3rd ed., McGraw-Hill, 2002.

R. Mailloux, *Phased Array Antenna Handbook*, 2nd ed., Artech House, 2005.

T. Milligan, *Modern Antenna Design*, John Wiley & Sons, 2005.

D. Miron, *Small Antenna Design*, Newnes Publishing, 2006.

C. Oestges and B. Clerckx, *MIMO Wireless Communications: From Real-World Propagation to Space-Time Code Design*, Academic Press, 2007.

A. Peterson et al., *Computational Methods for Electromagnetics*. IEEE Press, 1998.

T. Rappaport, *Wireless Communications,* 2nd ed, Prentice Hall, 2002.

S. Saunders and A. Aragon-Zavala, *Antennas and Propagation for Wireless Communication Systems,* 2nd ed., John Wiley & Sons, 2007.

C. Sletten, *Reflector and Lens Antennas*, Artech House, 1988.

J. Stratton, *Electromagnetic Theory*, McGraw-Hill, 1941.

W.L. Stutzman and G. Thiele, *Antenna Theory and Design,* 2nd ed., John Wiley & Sons, 1998.

Taflove and S. Hagness, *Computational Electrodynamics: The Finite Difference Time Domain Method,* 3rd ed., Artech House, 2005.

J. Volakis, *Antenna Engineering Handbook,* 4th ed., McGraw-Hill, 2007.

R. Waterhouse, *Printed Antennas for Wireless Communication*s, John Wiley & Sons, 2008.

Chapter 5 - *Facilities Infrastructure*

Tower Design: Minimum Design Loads for Buildings and Other Structures, SEI/ASCE Standard No. 7-05, Structural Engineering Institute.

E.P Carter, *Telecommunication Electrical Protection*, AT&T Technologies, 1985.

P. Hasse, *Overvoltage Protection of Low Voltage Systems*, 2nd. ed., IEE Power and Energy Series, 2000, ,

W. Reeve, *DC Power System Design for Telecommunications* Wiley-IEEE Press, 2007

Association Française de Normalisation (AFNOR), http://www.afnor.org/portail.asp?Lang=English

ATIS Protection Engineers Group, http://www.atis.org/peg/

British Standards Institution, http://bsi-global.com

Building Industry Consulting Service International, http://www/bicsi.org

Bureau of Indian Standards, http://www.bis.org.in/

CDMA Development Group, http://www.cdg.org

Federal Agency on Technical Regulating and Metrology (Russia), http://www.gost.ru/wps/portal/pages.en.Main

German Institute for Standardization (DIN), http://www.din.de/cmd?level=tpl-bereich&cmsareaid=47565&languageid=en

Japanese Standards Association, http://www.jsa.or.jp/default_english.asp

Low Voltage Directive (LVD), http://ec.europa.eu/enterprise/electr_equipment/lv/index.htm

National Lightning Detection Network, http://thunderstorm.vaisala.com/

NEMA Surge Protection Institute, http://www.nemasurge.com/

Standards Administration of China, http://www.standardsportal.org/splash/default.aspx

Standards Australia, http://www.standards.org.au/

Links for Power Standards in China:

http://blogs.zdnet.com/BTL/?p=3913

http://nbr.org/publications/specialreport/pdf/SR10.pdf

http://www.fuzing.com/qrx/CCN/941/power-standards

Links for Power Standards in Europe:

http://ieeexplore.ieee.org/xpl/RecentCon.jsp?punumber=4375

http://www.bksv.com/Applications/SoundPower/SoundPowerStandards.aspx

http://www.powercords.co.uk/standard.htm

Links for Power Standards in Japan:

http://www.iop.org/EJ/abstract/0026-1394/17/1/006

http://ieeexplore.ieee.org/Xplore/login.jsp?url=/iel1/39/10527/00491918.pdf?tp=&isnumber=&arnumber=491918

Links for Power Standards in other countries:

http://sciencelinks.jp/j-east/article/199914/000019991499A0308609.php

http://www.equitech.com/support/worldpwr.html

http://www.kropla.com/electric2.htm

Resources – Standards

Low Voltage Surge Protective Devices–Part 21: Surge Protective Devices Connected to Telecommunications and Signaling Networks–Performance Requirements and Testing Methods, AFNOR Groupe France Standard CEI 61643-21:2000.

Protection of Low-Voltage Electrical Installations against Overvoltages of Atmospheric Discharges and Switching–Selection and Erection of Surge Protective Devices, AFNOR Groupe France Standard UTE C15-443, August 2004,

Structural Standards for Steel Antenna Towers and Antenna Supporting Structures, ANSI/TIA-222.

Requirements for Electrical Installations–IEE Wiring Regulations, 17th ed., British Standards Institution BS7671:2008.

Application Guide for Electrical Relays for AC Systems, Bureau of Indian Standards IS 3842.

Electrical Equipment Designed for Use within Certain Voltage Limits (the "Low Voltage Directive"), European Commission Directive 2006/95/EC, August 2007.

Electrical Installations of Buildings, Section 444: Protection against Electromagnetic Interferences (EMI) in Installations of Buildings, Federal Agency on Technical Regulating and Metrology (Russia) GOST R50571.25-2001,.

Electrical Installations of Buildings, Section 442: Protection of Low Voltage Installations against Faults between High-Voltage Systems and Earth, German Institute for Standardization (DIN) VDE 0100-442:1997-11,.

Recommended Practice for Electric Power Distribution for Industrial Plant, IEEE Standard 141.

Recommended Practice for Grounding of Industrial and Commercial Power Systems, IEEE Standard 142.

Recommended Practice for Powering and Grounding Electronic Equipment, IEEE Standard 1100.

Electrical Installations of Buildings, Part 4-43: Protection for Safety–Protection against Overcurrent, JSA Standard JIS C 60364, 2006.

NFPA 70, 2008 NEC, Article 285: Surge Protective Devices (SPDs), 1 kV or Less.

NFPA 70, 2008 NEC, Article 800: Communications Circuits.

Electrical Installations (known as the "Australia/New Zealand Wiring Rules"), Standards Australia AS/NZ 3000, 2007.

Telcordia GR-1089-CORE, Issue 4, 2006, *Electromagnetic Compatibility and Electrical Safety–Generic Criteria for Network Telecommunications Equipment*.

Telcordia TR-NWT-0001011, Issue 1, 1992, *Generic Requirements for Surge Protective Devices (SPDs) on AC Power Circuits*.

Protectors for Paired-Conductor Communications Circuits, UL497, 7th ed., 2001.

Secondary Protectors for Communications Circuits, UL497A, 3rd ed., 2001.

Protectors for Data Communications and Fire-Alarm Circuits, UL497B, 4th ed., 2004.

Surge Protective Devices, UL1449, Issue 3, 2006.

Chapter 6 - Agreements, Standards, Policies and Regulations

ATIS NRSC: www.atis.org/nrsc

European Commission ARECI Study: www.bell-labs.com/ARECI

IEEE Communications Society: www.comsoc.org

IEEE CQR: www.comsoc.org/~cqr

IEEE Wireless Technical Committee: www.comsoc.org/socstr/org/operation/techcom/tcpc.html

ITU: www.itu.int

ITU-R: www.itu.int/ITU-R

International Wireless Association: www.ctia.org

NRIC: www.nric.org

U.S. FCC: www.fcc.org

WERT: www.wert-help.org

Chapter 7 – Fundamental Knowledge

[Bea00] H. W. Beaty, *Handbook of Electric Power Calculations*, McGraw-Hill, 2000.

[ETSI] http://www.etsi.org/WebSite/Standards/StandardsDownload.aspx.

[Gra02] C.F. Gray and E.W. Larson, *Project Management* , 2nd ed., McGraw-Hill, 2002.

[Hay04] J. Hayes and T.V.J. Ganesh Babu, *Modeling and Analysis of Telecommunications Networks*, Wiley-Interscience, 2004.

[IEEE CoE] "The IEEE Code of Ethics," available at http://www.ieee.org/portal/pages/iportals/aboutus/ethics/code.html.

IEEE Guide for Developing System Requirements Specifications, IEEE Standard 1233, 1998.

[ITU02] International Telecommunication Union, *Handbook of Satellite Communications*, John Wiley & Sons, 2002.

[Kha05] R.S. Khandpur, *Printed Circuit Boards*, McGraw-Hill, 2005.

[Kra99] J.D. Kraus and D. Fleisch, *Electromagnetics*, McGraw Hill, 1999.

[Lev79] K. Levitz and H. Levitz, *Logic and Boolean Algebra*, Barrons Educational Series Inc, 1979.

[Mill87] J. Millman and A. Grabel, *Microelectronics*, McGraw-Hill,1987.

[New04] D.G. Newnan, T.G. Eschenbach and J.P. Lavelle, *Engineering Economic Analysis*, 9th ed, Oxford University Press, 2004.

[New06] H. Newton, *Newton's Telecom Dictionary*, 22nd ed., Flatiron Publishing, 2006.

[Pau06] C.R. Paul, *Introduction to Electromagnetic Compatibility*, John Wiley & Sons, 2006.

[Pro07] J.G. Proakis and M. Salehi, *Digital Communications* McGraw-Hill Higher Education; 5th ed. (Nov, 2007).

[Tho06] R.E. Thomas, *The Analysis and Design of Linear Circuits*, John Wiley & Sons, 2006.

[Will06] W. Stallings, *Data and Computer Communications*, 8th ed., Prentice Hall, 2006.

[Wit02] R.A. Witte, *Electronic Test Instruments: Analog and Digital Measurements*, Prentice Hall, 2002.

[Whi03] J.F. White, *High Frequency Techniques: An Introduction to RF and Microwave Engineering*, John Wiley & Sons, 2003.

Appendix B

Creating the WEBOK

Development of Knowledge Areas for the Wireless Industry

Global communication is the defining political and economic force in the world today. It requires new ways of thinking and responding. For engineering professionals, recognizing and understanding this phenomenon is fast becoming a job requirement.

To address the worldwide wireless industry's growing and ever-evolving need for qualified communication professionals who can demonstrate practical problem-solving skills in real-world situations, the IEEE Communications Society (IEEE ComSoc) designed the IEEE Wireless Communication Engineering Technologies certification program. Individuals receiving this certification will be recognized as having the knowledge, skill, and ability to meet wireless challenges in various industry, business, corporate, and organizational settings.

The first task in the creation of any certification program is to identify the areas of expertise to be tested. To this end, ComSoc convened a meeting of 16 industry experts in New York City in December 2006. This group, called the Practice Analysis Task Force (PATF), was directed to draft a list of all the tasks and knowledge statements that a practicing wireless engineering professional with at least three years of experience must know. These statements were then grouped into seven areas called Domains, creating a document known as the Delineation.

The Delineation was then presented to 11 focus groups in seven cities in six countries. Nearly 90 volunteers examined each statement in the Delineation line by line. The deliberations of each focus group were recorded and each comment noted.

A second PATF meeting was held, once again in New York City, in May 2007. The consolidated information was reviewed, and suggested changes to the Delineation were carefully debated, resulting in a fine-tuned document.

The next step was to get feedback about this version of the Delineation from practicing wireless engineers. This was done with an Internet-driven survey delivered to over 5000 wireless engineers for a final validation of the Delineation. The survey asked each respondent to evaluate the task and knowledge statements in terms of:

- How often they encountered them in their career
- How important they are to the industry
- Who in their organization performed the tasks

236

- When in a practitioner's career would the knowledge be acquired
- How important was each task or item of knowledge to their career
- What fraction of their time did they personally spend on each task

Participants in the survey also provided extensive demographical information to enable ComSoc to assess the responses and further evaluate the viability of the Delineation. This discussion led to a final version of the Delineation.

Once the delineation of the body of knowledge was complete, the decision was made to use this same list not just for developing the certification exam but also as the basis for a Wireless Engineering Body of Knowledge. However, the WEBOK and the IEEE ComSoc Wireless Communication Engineering Technologies Certification program, though both based on the same knowledge areas, were created in completely separate efforts; the WEBOK presented here is a separate product unrelated to IEEE ComSoc WCET Certification.

Appendix C

Summary of Knowledge Areas

Chapter 1 — Wireless Access Technologies

Tasks:

1. Analyze multiple access schemes for various technologies.
2. Analyze spectrum implications in wireless access system design (examples might include applications, TDD/FDD, inter-modulation, LOS/NLOS, coverage/capacity).
3. Analyze design considerations and perform system design to eliminate coverage holes and to optimize capacity/coverage in urban/indoor areas.
4. Design a wireless access system (examples might include AP placement and channel selection) according to given bandwidth requirements, coverage, and other considerations.
5. Test devices with respect to interference issues in various operating environments (examples might include TDMA, CDMA, WCDMA, WLAN, 802.15).
6. Perform co-location interference analysis for systems (examples might include TDMA, CDMA, WCDMA, WLAN, 802.15, and GSM).
7. Compute the required bandwidth for a wireless system given certain network conditions (examples might include BER, flow count, and protocols in use).
8. Analyze the tradeoffs (examples might include bandwidth versus BER) of various error detection and correction techniques.
9. Analyze the tradeoffs (examples might include bandwidth versus power efficiency) and capacity implications of various power control schemes.
10. Calculate frequency re-use factor.
11. Design fundamental elements/attributes of wireless network systems (examples might include cellular, 802.16, WLAN, and satellite).
12. Analyze the steps involved in the process of handoff for various wireless systems (examples might include UMTS, CDMA2000, 802.16, and WLAN).

Knowledge of:

1. multiple access and multiplexing schemes (examples might include TDMA, CDMA, OFDMA, FDMA, and SDMA)
2. CDMA2000, 802.11, 802.15, and 802.16)
3. error detection and correction techniques
4. objectives of power-control schemes and their operation

5. handoff/mobility management
6. paging functions
7. the major components of a wireless network topology

Chapter 2 — Network and Service Architecture

Tasks:

1. Analyze service platforms including service enablers (examples might include messaging and positioning) and service creation/delivery (examples might include Open Service Access and Parlay).
2. Analyze IP addressing schemes for various technologies (examples might include Mobile IP, IPv4, and IPv6).
3. Design and test quality of service (QoS) (examples might include design and plan for adequate resources, selecting priority schemes, queuing strategies, and call administration control) for VoIP and IMS-based services.
4. Select and test a load-balancing scheme.
5. Analyze IP routing (examples might include interpreting an IP routing table).
6. Analyze ad hoc routing and mesh protocols, and suitability for various deployment scenarios.
7. Perform capacity planning using traffic engineering principles.
8. Perform error tracking and trace analysis on protocol control messages for specific systems.
9. Analyze the evolution of mobile networks to enable IP multimedia services (including circuit-switched to packet-switched network evolution).
10. Analyze intra- and inter-domain roaming.
11. Analyze the functioning of TCP/IP major transport protocols (examples might include TCP, UDP, and RTP) in the context of wireless communications.

Knowledge of:

1. IMS (IP multimedia subsystems) and its architecture, including session control and switching plane
2. VoIP/IP-multimedia protocols
3. wireless service enablers evolution
4. location and positioning techniques
5. load balancing principles in the context of wireless communications
6. IP routing and mobile IP networking and addressing schemes
7. error tracking and trace analysis techniques
8. circuit switched and packet switched data and packet cellular networks and the differences between them
9. roaming and roaming controls
10. TCP/IP including transport protocols

Chapter 3 — Network Management and Security

Tasks:

1. Design a fault monitoring system (examples might include using SNMP TRAP/NOTIFICATION).
2. Design a performance monitoring system (examples might include using SNMP GET/SET).
3. Develop/specify types and methods of alarm reporting for an installation.
4. Compute availability and reliability metrics from both the "network performance" and "system designer" perspectives (related to equipment failure).
5. Assess the potential impacts of known security attacks on wireless systems (examples might include virus, worm, DoS, and impersonation).
6. Plan corresponding solutions to known security attacks.
7. Monitor, log, and audit security-related data.
8. Analyze security vulnerabilities and prepare/recommend corrective actions.

Knowledge of:

1. quality of service (QoS) monitoring and control
2. fault management
3. configuration management
4. authentication, authorization, and accounting (AAA) principles and mechanisms
5. types of security attacks on wireless networks
6. protocols to secure wireless networks
7. security-violation events logging and monitoring
8. security issue management and resolution
9. network management protocols (examples might include simple network management protocol [SNMP])
10. performance metrics pertinent to various access networks
11. IP security, Encapsulation Security Payload (ESP), Internet Key Exchange, and digital signature
12. MIB, RMON, and Internet Control Messaging Protocol (ICMP)
13. Intrusion Detection Systems, DDoS Attacks, and traceback techniques

Chapter 4 — RF Engineering, Propagation, & Antennas

Tasks:

1. Calculate link budgets to evaluate system performance and reliability based on received signal level and fade margin (examples might include satellite, microwave link, base station to mobile station, wireless LAN, PAN, and free space optics).
2. Calculate path loss for various RF transmission systems (examples might include between isotropic or dipole reference antennas, base station to mobile station, base station to repeater,

earth station to satellite, LOS/NLOS paths, and clutter losses) and under varying atmospheric conditions (examples might include inversion layers, ducting, and variations in K factor).

3. Evaluate the effects of different fading models (examples might include Rayleigh and lognormal) and empirical path loss models on the received signal strength in various signal propagation environments (examples might include flat terrain, rolling hills, urbanized areas, and indoor environments [such as buildings or tunnels] with losses caused by walls, ceilings, and other obstructions).

4. Calculate and evaluate the effects on the received signal of path-related impairments, such as Fresnel Zone blockage, delay spread, and Doppler shift of a signal received by a moving receiver.

5. Calculate the polarization mismatch loss for various antenna systems (examples might include fixed microwave systems, cellular and mobile radio systems, and satellite systems).

6. Evaluate receive diversity gain for selection, equal gain, and maximal ratio diversity system configurations.

7. Determine parameters related to antennas or antenna arrays (examples might include pattern, beamwidth, gain, distance from an antenna or array at which far field conditions apply, spacing, beam forming, tilt, and sectorization) and analyze the effects of these parameters on coverage.

8. Determine appropriate antenna spacing at base station sites to prevent inter-system and intra-system interference effects, taking into account required radiation patterns and mutual coupling effects.

9. Generate and evaluate coverage and interference prediction maps for cellular, mobile radio, WLAN, and similar systems.

10. Develop a procedure to optimize the coverage of a radio system using propagation modeling and "drive test" measurements.

11. Develop a block diagram of an RF system (examples might include cellular, land mobile, and WLAN) employing standard modules (examples might include filters, couplers, circulators, and mixers) and/or use lumped or distributed matching networks, microstrips, and stripline.

12. Make RF system measurements (examples might include swept return loss to determine antenna system performance, transmitter output power [peak or average, as appropriate], signal-to-noise ratio at a receiver front end, and co-channel and adjacent channel interference for specific types of signal spectra).

Knowledge of:

1. different types of losses (examples might include transmission line loss, antenna gain, connector losses, and path loss)

2. procedures to calculate antenna gain and free space path loss

3. statistical fading models and distance-power (path loss) relationships in different propagation environments

4. the effects of outdoor terrain and indoor structures such as walls, floors, and ceilings on signal propagation

5. indoor and outdoor coverage calculation and verification techniques

6. Es/N0, EB/N0, RSSI, NF, and other system parameters

7. the relationship between receiver noise figure, noise temperature, and receiver sensitivity and the relationship between sensitivity under static conditions and the degradation of effective receiver sensitivity caused by signal fading in different propagation conditions

8. external noise sources and their impact on the S/N ratios of received signals, and of techniques for measuring the impact of external noise

9. basic antenna system design and use including antenna types (examples might include omnidirectional, panel, parabolic, dipole array, indoor antennas), antenna patterns, gain and ERP, antenna size, antenna polarization, receive and transmit diversity (examples might include MIMO) antenna systems, and proper antenna installation to provide for coverage, interference mitigation, and frequency reuse

10. adaptive antenna methods and techniques

11. subscriber unit, mobile, and device antennas and their performance characteristics

12. use of test equipment such as network analyzers, spectrum analyzers, and TDRs

13. co-channel and adjacent channel interference analysis and measurement methods and techniques

14. filters, power dividers, combiners, and directional couplers

Chapter 5 — Facilities Infrastructure

Tasks:

1. Determine the power consumption of a unit of communications equipment.
2. Determine the power consumption for a facility containing communications equipment.
3. Analyze the electrical protection requirements (includes grounding/earthing, bonding, shielding, and lightning protection) and design the electrical protection layout for a wireless telecommunications facility.
4. Determine the required antennas for the facility and their required positions on a structure.
5. Coordinate with other users when implementing a communications system in a shared location.
6. Develop a specification for the required structure for a wireless base station facility based on the required antenna sizes and elevations above ground.
7. Determine the required cable, antennas, and materials to implement an in-building wireless network.
8. Determine the required number of racks on which to mount the equipment and the rack layout and placement, taking into account the maintainability of the equipment.
9. Evaluate equipment compliance with industry standards, codes, and site requirements such as NEBS specifications, and ANSI, ETSI, IEC, and other applicable standards.

Knowledge of:

1. procedures to determine the power consumption of wireless communications equipment
2. how to determine the power required to support a site
3. the application of AC and DC power systems
4. the application of alternative energy sources to wireless communications facilities

5. heating, ventilation, and air-conditioning (HVAC) requirements
6. equipment racks, rack mounting spaces, and related hardware
7. electrical protection (including grounding/earthing, bonding, shielding, and lightning protection)
8. basic waveguides and transmission lines
9. tower specifications and standards
10. physical security requirements

Chapter 6 — Agreements, Standards, Policies, and Regulations

Tasks:

1. Assess service and equipment quality.
2. Prepare specifications for purchasing services and equipment, and evaluate the responses.
3. Verify compliance with regulatory requirements (examples might include licensing, standards, rules, and regulations).
4. Select and analyze frequency assignments.
5. Perform standardized homologation tests as required by regulatory or standardization bodies.
6. Evaluate compliance with health, safety, and environmental requirements.
7. Perform conformance/interoperability analyses of systems and components.
8. Analyze the use of licensed vs. unlicensed spectrum.
9. Obtain licenses and permits where required.

Knowledge of:

1. regulatory requirements (examples might include international, national, and local)
2. spectrum licensing
3. spectrum characteristics, availability, and management
4. local and site-specific rules and regulations
5. electrical safety (examples might include UL, EC, CSA, and IEEE C.95)
6. frequency assignment databases and online tools
7. modulation anomalies (examples might include cross modulation, modulation products, harmonics, and quantization impact)
8. health, safety, and environmental issues
9. equipment type approval processes/requirements

Chapter 7 — Fundamental Knowledge

Knowledge related to electrical engineering

1. fundamental AC/DC circuit analysis
2. mathematics including probability, statistics, and Boolean arithmetic

3. operation of complex test instruments, including oscilloscopes, spectrum analyzers, network analyzers, TDRs, and signal generators
4. Fourier frequency spectrum and transforms
5. basic printed circuit board design considerations
6. transmission theory and lines, antennas, basic optics, and basic electromagnetic wave theory and applications
7. power calculations (examples might include dB, dBm, and dBx)
8. basic concepts of queuing theory and traffic analysis
9. basic signal processing (examples might include analog, digital, and statistical)
10. basic concepts related to optical communications
11. basic electronic system-level block diagrams
12. basic power supply design

Knowledge related to communication systems

1. basic communication and information theory (analog and digital)
2. basic telephony (including signaling, switching, and transmission)
3. noise impairments
4. basic EMI, EMC, and interference
5. frequency allocations and reuse
6. how to identify and locate appropriate industry technical standards, codes, and other applicable requirements
7. modulation techniques for analog (examples might include AM, FM, and PM)
8. modulation techniques for digital (examples might include FSK, PSK, and QAM)
9. wireless multiple-access schemes (examples might include FDMA, TDMA, CDMA, and variants)
10. basic satellite communications
11. digital data transmission formats (examples might include E1/T1 and OC-n/SDH)
12. basic components of RF circuitry
13. basic RF circuit design
14. basic RF coupling, radiation, and antenna theory concepts
15. measurements for RF circuits and sub systems, such as output power, receiver sensitivity, noise figure, linearity performance, and spectral performance

Knowledge of general engineering management:

1. project management methods and processes
2. fundamental engineering economics
3. design and configuration for ease of maintenance
4. documentation and configuration control schemes
5. IEEE Code of Ethics

Appendix D

Glossary

3DES	Encryption Standard
3G	Metropolitan Area Network
3GPP	3rd Generation Partnership Project
3GPP2	3G Partnership Project 2
A5	Encryption algorithm
AAA	Authentication Authorization Accounting
AAD	Additional Authentication Data
ACK	Acknowledge
ACM	Address Complete Message
ADC	Analog to Digital Converted
AES	Advanced Encryption Standard
AF	Diffserv Assured Forwarding
AGC	Automatic Gain Control
AMC	Adaptive Modulation and Coding
AMPS	Advanced Mobile Phone System
ANM	Answer Message
ANSI	American National Standards Institute
AR	Axial Ratio for Elliptical Polarization
ARIB	Association of Radio Industries and Business
AS	Application Server
ASCII	American Standard Code for Information Interchange
ASN	Access Service Network

ASN.1	Abstract Syntax Notation One
ASP	Application Service Provider
ATIS	Association Telecommunications Industries Standards
ATM	Asynchronous Transfer Mode
AuC	Authentication Center
AUT	Antenna Under Test
AUTN	Network authentication token
AUTS	Token used in resynchronization
AWS	Advanced Wireless Services
BCMCS	Broadcast and Multicast Services
BE	Best Effort
BSC	Base Station Controller
BSS	Basic Service Set
BTS	Base Transceiver Station
CBC	Cipher Block Chaining Message
CBC-MAC	Cipher Block Chaining Message Authentication Code
CC	Call Control
CCCH/BCCH	Common Control Channel Broadcast Control Channel
CCI	Co Channel Interference
CCM	CTR Mode with CBC-MAC

CCMP	Counter Mode with Cipher Block Chaining Message Authentication Code Protocol
CCSA	China Communications Standard Association
CDMA	Code Division Multiple Access
CGM	Conjugate Gradient Method
CID	Connection ID
CIR	Carrier to Interference Ratio
CM	Connection Management
CMA	Constant Modulus Algorithm
COMP128	Algorithm
CP	Cyclic Prefix
CP	Circular Polarization
CQI	Channel Quality Indicator
CRC	Cyclic Redundancy Check
CRC-32	Cyclic Redundancy check 32 bits
CS	Coding Scheme
CSCF	Call Session Control Function
CSMA/CA	Carrier Sense Multiple Access with Collision Avoidance
CSN	Connectivity Service Network
CST	Computer Simulation Technology
CTIA	International Association for the Wireless Telecommunications Industry
CTS	Clear To Send
DARPA	Defense Advanced Research Projects
dBi	Decibel Isotropic
dBm	Decibel milliwatts
dBr	Decibel Relative
DCH	Dedicated Channel
DECT	Digital Enhanced Cordless Telephony
DES	Data Encryption Standard
DiffServ	Packet Classification

DIFS	Distributed Inter Frame Space
DL	Down Link
DMB	Digital Multimedia Broadcasting
DNS	Domain Name System
DRA	Dielectric Resonator Antenna
DRC	Data Rate Control
DSL	Digital Subscriber Line
DSS1	Digital Subscriber Signaling System 1
DVB-H	Digital Video Broadcast Handheld
EAP	Extensible Authentication Protocol
EAP-FAST	EAP Flexible Authentication via Secure Tunneling
EAPoL	EAP Over LAN
EAP-TLS	EAP Transport Layer Security
EAP-TTLS	EAP Tunneled TLS
E-DCH	Enhanced Dedicated Channel
EDGE	Enhanced Data Rates for GSM Evolution
EF	Diffserv Expedited Forwarding
EGPRS	Enhanced GPRS
EIA	Electronic Industries Alliance
EIRP	Effective Isotropic Radiated Power
EM	Electromagnetic
EP	Elliptical Polarization
ERP	Effective Radiated Power
ESS	Extended Service Set
eTOM	Enhanced Telecom Operations Map
ETSI	European Telecommunications Standard Institute
FA	Foreign Agent
FACA	U.S. Federal Advisory Committee Act
FBSS	Fast Base Station Switching

FCC	Federal Communications Commission		HSDPA	High Speed Down Link Packet Access
FCC	Federal Communications Commission		HS-DSCH	High Speed Downlink Shared Channel
FDD	Frequency Division Duplex		HSPA	High Speed Packet Access
FDMA	Frequency Division Multiple Access		HSS	Home Subscriber Server
FDMA	Frequency Division Multiple Access		HSUPA	Enhanced Uplink E-DCH
FDTD	Finite Difference Time Domain		HTTP	Hypertext Transfer Protocol
FEM	Finite Element Method		IBSS	Independent Basic Service Set
FFT	Fast Fourier Transform		I-CSCF	Interrogating CSCF
FSO	Free Space Optics		ICV	Integrity Check Value
FSS	Frequency Selective Surfaces		IDEN	Integrated Digital Enhanced Network
G.711	Encoder		IEC	International Electro Technical Commission
GGSN	Gateway GPRS Support Node		IETF	Internet Engineering Task Force
GKH	Group Key Hierarchy		IF	Intermediate Frequency
GMSC	Gateway Mobile Switching Centre		IFFT	Inverse Fast Fourier Transform
GPRS	General Packet Radio Service		IK	Integrity Key
GPRS	General Packet Radio Service		IKE	Internet Key Exchange
GPS	Global Positioning System		IMS	IP Multimedia System
GSM	Global System for Mobile Communications		IMSI	International Mobile Subscriber Identity
GTC	Generic Token Card		IMT-2000	International Mobile Telecommunications 2000 ITU standard
H.263	Video Codec Low-Bit rate		IP	Internet Protocol
H.264	Video Codec MPEG-4 Advanced Video Codec		IP v4	Internet Protocol Version 4
HARQ	Hybrid Automatic Repeat Request		IP v6	Internet Protocol Version 6
HARQ	Home Agent		IP-CAN	IP Connectivity Access Networks
HE	Home Environment		IPSec	Protocols for Security
HFSS	High Frequency Structure Simulator		IS-136	Interim Standard 136
HHO	Hard Handoff		IS-95	Interim Standard 95 CDMA ONE
Hi-Cap	High Capacity		ISM	Industry Science and Medical RF Band
HLR	Home Location Register		ISO	International Standard Organization
HLR/AUC	Home Location Register/Authentication Center		ISUP	ISDN User Part
HN	Home Network		ISUP IAM	ISUP Initial Address Message
HO	Handoff			

ITIL	Information Technology` Infrastructure Library		MIB	Management Information Base
ITU	International Telecommunication Union		MIC	Message Integrity Code
ITU-R	International Telecommunication Union–Radio		MIMO	Multiple Input Multiple Output
KA	Knowledge Area		MIP	Mobile IP
KC	Ciphering Key		MISO	Multiple Input Single Output
KCK	EAPoL Key Communication Key		MM	Mobility Management
KEK	EAPoL Key Encryption Key		MMUSIC	Multiparty Multimedia Session Control
LAN	Local Area Network		MoM	Method of Moments
LDPC	Turbo Code		MOS	Mean Opinion Score
LH	Left Hand Circular Polarization		MPDU	MAC Protocol Data Unit
LMS	Least Mean Square		MPEG	Moving Picture Expert Group
Lo-Cap	Low Capacity		MPLS	Multiprotocol Label Switching
LOS	Line of Sight		MRF	Media Resource Function
LOS	Local Oscillator		MS	Mobile Station
LP	Linear Polarization		MSC	Mobile Switching Center
LS-CMA	Least Squares Constant Modulus Algorithm		MSC/VLR	Mobile Switching Center/Visitor Location Register
LTE	Long Tern Evolution		MU-MIMO	Multiple User MIMO
LTE	Long Term Evolution		NACK	Not Acknowledge
MAC	Media Access Control		NAS	Network Access Server
MAC	Message Authentication Code		NAV	Network Allocation Vector
MAC-S	Authentication token used in resynchronization		NEBS	Network Equipment Building Systems standard
MAN	Metropolitan Area Network		NEC	Numerical Electromagnetic Code
MAP	Mobile Application Part		NF	Noise Figure
MBMS	Multimedia Broadcast/Multicast Service		NFC	Near Field Communication
MCW	Multi Codeword		NGMC	Next Generation Mobile Committee
MD5	Message Digest 5		NGMN	Next Generation Mobile Networks
MDHO	Macro Diversity Handover		NGN	New Generation Network
MDS	Minimum Discernible Signal		NIC	Network Interface Card
MEdiaFLO	Forward Link Only Technology		NIST	National institute of Standards and Technology
MGCF	Media Gateway Control Function		NLOS	Non Line of Sight
MGW	Media Gateway		NMHA	Normal Mode Helical Antenna

NRSC	Network Reliability Steering Committee		PSTN	Public Switched Telephone Network
NSP	Network Service Provider		QAM	Quadrature Amplitude Modulation
NSS	Network Subsystem		QoS	Quality of Service
NSTAC	National Security Telecommunications Advisory Committee (U.S.)		QPSK	Quadrature Phase Shift Keying
OATS	Open Area Test Site		RACH	Random Access Channel
OFDMA	Orthogonal Frequency Division Multiple Access		RADIUS	Remote Access Dial In User Server
OGC	Office of Government Commerce		RAN	Radio Access Network
OSA	Opportunistic Spectrum Address		RAND	Random
OSI	Open Systems Interconnect		RC4	RC4 Cipher Algorithm
OSPF	Open Shortest Path First Routing Protocol		RET	Remote Electrical Tilt
OSS/BSS	Operational and Business Support Systems		RF	Radio Frequency
OTA	Over The Air Programming		RFC	Request For Comment
OTP	One Time password		RFC	Request for Change
PAN	Personal Area Network		RFID	Radio Frequency Identification
PAPR	High Peak to Average Power Ratio		RHCP	Right Hand Circular Polarization
P-CSCF	Proxy CSCF		RLS	Recursive Least Squares
PDC	Personal Digital Cellular		RNC	Radio Network Controller
PDSN	Packet Data Serving Node		ROAMOPS	IETF Roaming Operations
PDSN	Packet Data serving Node		ROHC	Robust Header Compression
PDU	Protocol Data Unit		RR	Radio Resource
PEAP	Protected EAP		RRC	Radio Resource Control
PFDM	Orthogonal Frequency Division Multiplex		RSA	Rivest Shamir Alderman
PHY	Physical		RSN	Robust Security Networks
PIFA	Planar Inverted F Antenna		RSNA	Robust Security Network Associations
PIN	Personal Identification Number		RTP	Real Time Protocol
PKH	Pairwise Key Hierarchy		RTS	Request to Send
PL	Path Loss		RTT	Round Trip Time
PLMN	Public Land Mobile Networks		S/N	Signal to Noise Ratio
PN	Pseudo Noise		SA	Security Association
PO	Physical Optics		SCP	ETSI Smart Card Platform
PPP	Point to Point Protocol		S-CSCF	Serving CSCF
PSK	Phase Shift Keying		SCTP	Stream Control Transmission Protocol
PSTN	Public Switched Telephone Network		SCW	Single Codeword
			SDCCH	Stand Alone Dedicated Channel
			SDMA	Space Division Multiple Access
			SDR	Software Defined Radio

SEGF	Security Gateway Function		TCP	Transmission Control Protocol
SF	Spreading Factor		TCP/IP	Suite of Protocols
SFDR	Spurious Free Dynamic Range		TD-CDMA	Time Division CDMA
SFID	Service Flow ID		TDD-HCR	Time Division Duplex
SGSN	Serving GPRS Support Node		TDMA	Time Division Multiple Access
SGW	Signaling Gateway		TD-SCDMA	Time Division Synchronous CDMA
SID	System Identification Number		TIA	Telecommunications Industry Association
SIG	Special Interest Group of WWRF		TK	Temporal Key
SIM	Subscriber Identity Module		TKIP	Temporary Key Integrity Protocol
SIMO	Single Input Multiple Output		TMF	TM Forum
SIP	Session Initiation Protocol		TS	Time Slot
SIR	Signal to Interference Ratio		TSC	TKIP Sequence Counter
SISO	Single Input Single Output		TSG CT	TSG Core Network & Terminals
SLF	Subscriber Location Function		TSG GERAN	TSG GSM EDGE Radio Access Network
SMI	Structure of Management Information		TSG RAN	TSG Radio Access Network
SMS	Short Message Service		TSG SA	ETG Services & System Aspects
SM-SC	Short Message Service Center		TTA	Telecommunications Technology Association of Korea
SMTP	Simple Message Transfer Protocol		TTC	Telecommunications Technology Committee
SNMP	Simple Network Management Protocol		UDP	User Datagram Protocol
SPC	Single Parity Check		UE	User Equipment
SQN	Sequence		UMB	Ultra Mobile Broadband
SRES	Signed Response		UMTS	Universal Mobile Telecommunications System
SRTP	Secure RTP		UMTS AKA	Protocol used in 3G
SS7	Signaling System 7		URA	Universal Terrestrial Radio Access
SSB	Single Sideband		USGS	United States Geological Survey
SSID	Service Set Identifier		USIM	UMTS SIM
STA	Stations		UTRA	Universal Terrestrial Radio Access
T2P	Traffic To Pilot		UTRA TDD-HCR	TD-CDMA UTRA MODE
TCAP	Transaction Capabilities application Part		UTRA TDD-LCR	TD-SCDMA UTRA MODE
TCH	Traffic Channel			
TCH/FS	Traffic Channel Full Rate Speech			
TCH/HS	Traffic Channel Half Rate			

UTRAN	UMTS Terrestrial Radio Access Network
UWB	Ultra Wideband
VLR	Visitor Location Register
VN	Visited Network
VoIP	Voice Over IP
VSWR	Voltage Standing Wave Ratio
WAN	Wide Area Network
W-CDMA	Wideband CDMA
WEP	Wireless Encryption Protocol
WERT	Wireless Emergency Response Team
WG	Working Group of WWRF
WiFi	Wireless Fidelity
WiMAX	Worldwide Interoperability for

	Microwave Access
WINNER	Wireless World Initiative New radio
WLAN	Wireless LAN
WMN	Wireless Mesh Network
WPA2	Wi-Fi Protected Access
WRC	World Radio Communication Conference
WWRF	Wireless World Research Forum
XG	Next Generation
XMAC (PG 26)	Cryptographic primitive in the 3GSM Key Generation Process
XOR	Exclusive Or

Appendix E

About the IEEE Communications Society

The IEEE Communications Society is a diverse group of industry professionals with a common interest in advancing all communication technologies. Individuals within this unique community interact across international and technological borders to produce publications, organize conferences, foster educational programs, promote local activities, and work on technical committees

Website: www.comsoc.org

Conferences

Every year, the IEEE Communications Society sponsors major conferences that attract hundreds of the best quality paper/presentation submissions and attendees. Held at convenient locations around the world, these meetings attract thousands of participants who have much to share beyond their strong desire to learn.

Communications Society conferences and workshops provide ideal opportunities to be part of the latest technological developments, and network with leaders who are changing the world of communications.

- IEEE/OSA Conference on Optical Fiber Communications/National Fiber Optic Engineers Conference
- (OFC/NFOEC)
- IEEE Wireless Communications and Networking Conference (WCNC)
- IEEE/IFIP Network Operations & Management Symposium (NOMS)
- IFIP/IEEE International Symposium on Integrated Network Management (IM)
- IEEE International Conference on Communications (ICC)
- IEEE International Enterprise Networking & Services Conference (ENTNET)
- IEEE Conference on Computer Communications (INFOCOM)
- IEEE/AFCEA Military Communications Conference (MILCOM)
- IEEE Global Communications Conference (GLOBECOM)
- IEEE Consumer Communications & Networking Conference (CCNC)
- IEEE International Symposium on Personal, Indoor and Mobile Radio Communications (PIMRC)
- IEEE Conference on Sensor and Ad Hoc Communications and Networks (SECON)
- IEEE International Symposium on Dynamic Spectrum Access Networks (DySPAN)
- IEEE International Symposium on Power Line Communications and Its Applications (ISPLC)

Publications

- *IEEE Communications Magazine*
- *IEEE Network: The Magazine of Global Internetworking*

- *IEEE Wireless Communications*
- *IEEE Communications Letters*
- *IEEE Transactions on Communications*
- *IEEE Journal on Selected Areas in Communications*
- *IEEE/ACM Transactions on Networking*
- *IEEE Transactions on Wireless Communications*
- *Transactions on Network and Service Management (TNSM)*
- *IEEE Communications Surveys & Tutorials*
- *IEEE Transactions on Mobile Computing*
- *IEEE/OSA Journal of Lightwave Technology*
- *ComSoc e-News*
- *ComSoc Digital Library*